Parasites and Parasitic Diseases in Small Animals

Parasites and Parasitic Diseases in Small Animals

Guest Editors

Angela M. García-Sánchez
Rocio Callejón

MDPI

Basel • Beijing • Wuhan • Barcelona • Belgrade • Novi Sad • Cluj • Manchester

Guest Editors
Angela M. García-Sánchez
Department of Microbiology
and Parasitology
Faculty of Pharmacy
University of Seville
Sevilla
Spain

Rocio Callejón
Department of Microbiology
and Parasitology
Faculty of Pharmacy
University of Seville
Sevilla
Spain

Editorial Office
MDPI AG
Grosspeteranlage 5
4052 Basel, Switzerland

This is a reprint of the Special Issue, published open access by the journal *Animals* (ISSN 2076-2615), freely accessible at: https://www.mdpi.com/journal/animals/special_issues/C71ECR8KGT.

For citation purposes, cite each article independently as indicated on the article page online and as indicated below:

Lastname, A.A.; Lastname, B.B. Article Title. *Journal Name* **Year**, *Volume Number*, Page Range.

ISBN 978-3-7258-4439-5 (Hbk)
ISBN 978-3-7258-4440-1 (PDF)
https://doi.org/10.3390/books978-3-7258-4440-1

Contents

About the Editors

Angela M. García-Sánchez

Angela M. García-Sánchez holds a degree in Pharmacy (2010) from the University of Seville, where she also completed an M.Sc. in Biotechnology and Molecular Biology (2011) and a Ph.D. in Microbiology (2016). She is currently an Assistant Professor in the Department of Microbiology and Parasitology at the University of Seville. Angela has contributed to eight scientific research projects and has authored over 20 publications in leading journals and conferences in the field of Microbiology and Parasitology. Her current research focuses on parasite morphometrics and the complex interactions between hosts, parasites, and microbiota.

Rocio Callejón

Rocío Callejón obtained her Pharmacy degree in 2007 and a Master's in Molecular Genetics and Microbial Biotechnology in 2008. She completed her PhD with a predoctoral grant from the University of Seville and the FPU program, focusing on classical and molecular taxonomy, phylogenetic studies, and diagnostic techniques for *Trichuris* nematodes. Her research has contributed significantly to integrative taxonomy, combining morphobiometric, molecular, and proteomic data, and has led to key phylogeographic analyses of *Trichuris* species populations in rodents. Callejón's work has also advanced diagnostic techniques for *Trichuriasis,* including the MALDI-TOF method, and explored therapeutic applications for autoimmune diseases, such as Crohn's disease. She has published 23 articles in high-impact journals and won the Avenzoar Pharmaceutical Foundation Research Award in 2015. She currently co-supervises a doctoral thesis and has contributed to more than nine research projects, with collaborations in Argentina, Italy, Iran, and Denmark. Since 2013, she has been researching the molecular taxonomy of *Siphonaptera,* a group with significant public health implications. Her scientific output includes 34 articles in JCR journals and an h-index of 20.

animals

Editorial

Parasites and Parasitic Diseases in Small Animals

Angela M. García-Sánchez * and Rocío Callejón

Department of Microbiology and Parasitology, Faculty of Pharmacy, University of Seville, Profesor García González 2, 41012 Sevilla, Spain; callejon@us.es
* Correspondence: agarcia77@us.es

1. Introduction

Parasites and parasitic diseases in small animals pose a significant challenge to public health, animal health, and environmental sustainability, and this Special Issue is particularly relevant within the context of the One Health approach, which recognizes their interconnectedness [1].

Humans coexist with animals that are part of our ecosystem, although they are often not visible due to their small size. In addition to animals naturally present in the environment, a large number of small mammals are commonly kept as pets—including dogs, cats, rodents, hedgehogs, and rabbits, not to mention other exotic animals like birds, reptiles, amphibians, and fish. These close relationships explain why some parasitic diseases are also zoonotic, infecting humans and even causing severe illnesses. Zoonoses can also lead to significant economic losses due to decreased productivity and increased healthcare costs. Domestic and wild animals can act as reservoirs and vectors for zoonotic parasites, facilitating disease transmission between species and increasing the risk of outbreaks that compromise both human well-being and the ecological balance [1,2].

Flea infestations and many other parasitic infections are examples that illustrate how parasites can affect multiple hosts in various environments [3–5]. These diseases not only cause severe health problems in humans and animals but also lead to significant economic losses due to treatment and control costs, as well as decreased productivity in sectors related to livestock and wildlife management [6]. Furthermore, factors such as climate change, urbanization, and increased human–animal contact have exacerbated the spread of parasites to new geographical regions [6,7].

Although there have been advances in this field in recent years, many aspects remain unknown. The success of the One Health concept now requires breaking down the interdisciplinary barriers that still separate human and veterinary medicine from ecological, evolutionary, and environmental sciences. The surveillance and control of these diseases is essential to mitigate their impact and, without effective strategies that limit transmission from reservoir animals to humans, the optimal control of these infections is not possible. The implementation of measures based on the One Health approach encourages interdisciplinary collaboration among physicians, veterinarians, ecologists, and other specialists to develop integrated solutions that address both the underlying causes and consequences of zoonotic parasitoses [8,9].

In this regard, it is crucial to strengthen epidemiological monitoring initiatives, improve diagnostic tools, and promote educational programs targeted at the general public and health professionals. These actions will not only protect human and animal health but also contribute to maintaining healthy and sustainable ecosystems in an increasingly interconnected world [8,10]. For these reasons, studying and controlling parasitic diseases in small animals is crucial to improving global public health.

Animals **2025**, *15*, 1183 https://doi.org/10.3390/ani15081183

2. The Structure of This Special Issue

This Special Issue includes eleven papers addressing the challenge of parasitic diseases in small animals, exploring their impact on both wildlife and domestic species. It includes studies on the diversity and host specificity of avian haemosporidians in an Afrotropical conservation region, suggesting that biodiverse areas may harbor a greater variety of parasites, which is crucial for understanding host–parasite dynamics and potential disease emergence in avian populations. On the other hand, the application of geometric morphometrics is introduced to differentiate three populations of synanthropic fleas in Andalusia (Spain), highlighting the importance of this technique for studying and managing arthropod communities.

Further research investigates the detection of β-tubulin polymorphisms in *Trichuris trichiura*, offering promising tools for improving the diagnosis and treatment of trichuriasis. In addition, the prevalence and risk factors of gastrointestinal parasites in domestic dogs in Serbia are explored, providing recommendations to improve parasite control in canines. The Special Issue also covers the occurrence of *Platynosomum illiciens* infection in cats with elevated liver enzymes, shedding light on the epidemiology of this parasitic infection and its effects on feline health.

Additionally, one study examines the status of *Trypanosoma grosi* and *Babesia microti* in small mammals in the Republic of Korea, emphasizing the need to understand the distribution and genetic diversity of these parasites to better manage zoonotic diseases; another reports the detection of *Theileria sinensis*-like and *Anaplasma capra* in ticks, contributing to the knowledge of the distribution and genetic diversity of these pathogens.

The prevalence of *Sarcocystis* spp. macrocysts in wildfowl in the Eastern Baltic region is analyzed over a specific time frame, with a focus on the economic impact of *Sarcocystis rileyi* infections on game birds. Data on new intermediate and accidental hosts of *Angiostrongylus cantonensis* in the Canary Islands (Spain) highlight the parasite's spread into new areas, underscoring the importance of monitoring its distribution. In addition, the prevalence of microsporidia in the North African hedgehog (*Atelerix algirus*) in the Canary Islands is reported, and its implications for wildlife health and conservation are discussed. Finally, the detection of *Sarcocystis pilosa* in a red fox (*Vulpes vulpes*) reintroduced in South Korea provides insights into the presence of intermediate and definitive hosts in the region, emphasizing the need for monitoring reintroduced wildlife for parasitic infections.

These studies collectively illustrate the complex relationships between parasitic infections, animal health, and ecosystem dynamics.

3. Conclusions

This Special Issue highlights a variety of innovative contributions that provide a better understanding of the epidemiology, host specificity, and diversity of parasites and parasitic diseases in small animals. Monitoring and controlling these diseases not only protects animal health but also safeguards public health by reducing the risk of zoonotic infections, due to their close contact with humans, without forgetting the potential outbreaks that may arise in the near future.

Author Contributions: All authors listed have made a substantial, direct, and intellectual contribution to the work and have approved it for publication. All authors have read and agreed to the published version of the manuscript.

Acknowledgments: The Guest Editors of this Special Issue thank all of the authors for their important contributions.

Conflicts of Interest: The authors declare no conflicts of interest.

References

1. Krecek, R.C.; Rabinowitz, P.M.; Conrad, P.A. Demystifying and Demonstrating the Value of a One Health Approach to Parasitological Challenges. *Vet. Parasitol.* **2020**, *287*, 109202. [CrossRef] [PubMed]
2. Webster, J.P.; Gower, C.M.; Knowles, S.C.; Molyneux, D.H.; Fenton, A. One Health—An Ecological and Evolutionary Framework for Tackling Neglected Zoonotic Diseases. *Evol. Appl.* **2016**, *9*, 313–333. [CrossRef] [PubMed]
3. Cable, J.; Barber, I.; Boag, B.; Ellison, A.R.; Morgan, E.R.; Murray, K.; Pascoe, E.L.; Sait Steven, M.; Wilson, A.J.; Booth, M. Global change, parasite transmission and disease control: Lessons from ecology. *Phil. Trans. R. Soc.* **2017**, *372*, B37220160088. [CrossRef] [PubMed]
4. Betson, M.; Alonte, A.J.I.; Ancog, R.C.; Aquino, A.M.O.; Belizario, V.Y., Jr.; Bordado, A.M.D.; Clark, J.; Corales, M.C.G.; Dacuma, M.G.; Divina, B.P.; et al. Zoonotic Transmission of Intestinal Helminths in Southeast Asia: Implications for Control and Elimination. *Adv. Parasitol.* **2020**, *108*, 47–131. [CrossRef] [PubMed]
5. Morelli, S.; Diakou, A.; Di Cesare, A.; Colombo, M.; Traversa, D. Canine and Feline Parasitology: Analogies, Differences, and Relevance for Human Health. *Clin. Microbiol. Rev.* **2021**, *34*, e0026620. [CrossRef] [PubMed]
6. Kaminsky, R.; Mäser, P. Global impact of parasitic infections and the importance of parasite control. *Front. Parasitol.* **2025**, *4*, 1546195. [CrossRef] [PubMed]
7. Mott, K.E.; Desjeux, P.; Moncayo, A.; Ranque, P.; de Raadt, P. Parasitic diseases and urban development. *Bull. World Health Organ.* **1990**, *68*, 691–698. [PubMed]
8. Erkyihun, G.A.; Alemayehu, M.B. One Health approach for the control of zoonotic diseases. *Zoonoses* **2022**, *2*, 963. [CrossRef]
9. Edward, M.; Heniedy, A.M.; Saminu, A.; Florence Mary, J.J.; Ahmed, D.A.; Engmann, S.T.; Onyeaghala, C.; Shah, S. Climate change and contagion: The emerging threat of zoonotic diseases in Africa. *Infect. Ecol. Epidemiol.* **2025**, *15*, 2441534. [CrossRef] [PubMed]
10. Hao, Y.W.; Wang, Q.; Cao, C.L.; Tian, T.; Zhu, Z.L.; Xu, J.; Zhou, S.; Wu, W.; Chen, Y.; Zhang, Y.; et al. Construction and application of surveillance and response systems for parasitic diseases in China, led by NIPD-CTDR. *Adv. Parasitol.* **2020**, *110*, 349–371. [CrossRef] [PubMed]

animals

MDPI

Article

Diversity and Host Specificity of Avian Haemosporidians in an Afrotropical Conservation Region

Mduduzi Ndlovu [1,*], Maliki B. Wardjomto [1], Tinotendashe Pori [2] and Tshifhiwa C. Nangammbi [3]

1 School of Biology and Environmental Sciences, University of Mpumalanga, Mbombela 1201, South Africa
2 School of Life Sciences, University of Warwick, Coventry CV4 7AL, UK
3 Department of Nature Conservation, Tshwane University of Technology, Pretoria 0001, South Africa
* Correspondence: mduduzindlovu@gmail.com

Simple Summary: African tropical regions have a remarkably high bird diversity, yet few studies have tried to unravel the presence of blood parasites in birds found in conservation areas. Knowing which blood parasites are present will help us to prepare for potential disease outbreaks. We test the hypothesis that conservation regions have a high diversity of parasites. Molecular methods were used to screen 1035 blood samples from 55 bird species for blood infections on sites inside and adjacent to the Kruger National Park in South Africa. Overall, 28.41% of birds were found infected with at least one type of blood parasites. Bird malaria of the type *Haemoproteus* and *Plasmodium* was found in 17.39% and 4.64% of the birds respectively. *Leucocytozoon* blood parasite was found in 9.24% of birds. One hundred distinct blood parasite types were detected, of which 56 were new types. Similar bird malaria (*Haemoproteus* and *Plasmodium*) infections were found in closely related birds, while *Leucocytozoon* was found in almost every bird type. Sites with a high bird diversity also had a high parasite diversity. These findings provide insight of how birds can gradually survive their blood parasite infections.

Abstract: Afrotropical regions have high bird diversity, yet few studies have attempted to unravel the prevalence of avian haemosporidia in conservation areas. The diversity and host specificity of parasites in biodiversity hotspots is crucial to understanding parasite distribution and potential disease emergence. We test the hypothesis that biodiverse regions are associated with highly diverse parasites. By targeting the cytochrome b (*Cytb*) gene, we molecularly screened 1035 blood samples from 55 bird species for avian haemosporidia infections to determine its prevalence and diversity on sites inside and adjacent to the Kruger National Park. Overall infection prevalence was 28.41%. *Haemoproteus*, *Leucocytozoon*, and *Plasmodium* presented prevalences of 17.39%, 9.24%, and 4.64%, respectively. One hundred distinct parasite lineages were detected, of which 56 were new lineages. *Haemoproteus* also presented the highest diversity compared to *Leucocytozoon* and *Plasmodium* with varying levels of specificity. *Haemoproteus* lineages were found to be specialists while *Plasmodium* and *Leucocytozoon* lineages were generalists. We also found a positive relationship between avian host diversity and parasite diversity, supporting an amplification effect. These findings provide insight data for host–parasite and co-evolutionary relationship models.

Keywords: avian malaria; diseases; diversity; parasites; prevalence

1. Introduction

The diversity of parasites is an important selective force shaping communities and ecosystems. Parasites generally have a higher mutation rate than their hosts, in order to evade the host immune system and successfully proliferate in an ecosystem [1]. The parasites' ability to diversify and mutate can lead to niche expansions and pathogen host shifts allowing the infection of numerous hosts [2]. On the other hand, parasite diversity may be driven by the diversity and evolutionary life-history traits of the available hosts [3] as well as the prevailing environmental factors [4]. Changes in the environment affect hosts as well as the host–parasite relationship [5]. Host specificity is therefore a strong determinant of both parasite diversity and prevalence, which provides an opportunity to understand the mechanisms driving parasite spillovers and factors linked with emerging infectious diseases [6].

The survival and proliferation of parasites in an ecosystem is dependent on several evolutionary strategies and traits developed in relation to their host (The Red Queen hypothesis) and the habitat type where they occur. One such trait is host specificity, which represents the number and/or diversity of the host species that a parasite can infect [7]. Such a strategy can be explained by the "Trade-off" hypothesis and the "Niche-breadth" hypothesis [8]. The parasite could be a generalist, i.e., it can infect multiple host species achieving a high or low prevalence, as opposed to a specialist that will be found in only one or a few closely related species, achieving a higher prevalence than a generalist by predominantly infecting more closely related host species [9,10]. The ability of a parasite to infect a host is also dependent on the presence of susceptible hosts, competent vectors, and a permissive environment [11]. As such, a parasite that is a specialist in one ecosystem may appear to be a generalist in another and vice versa [10], a pattern that was observed with vector species as well [12]. Host switching strategies may also occur where parasites infect many host species to avoid the host's defences [13]. In birds, a host species' anti-parasite behaviours (i.e., body maintenance, nest maintenance, avoidance of parasitized prey using cues to their presence in conspecifics and intermediate hosts, migration, and tolerance) in combination with immune system defences may reduce a parasite load thus driving host switching by the parasite [14]. Although poorly understood, the interaction of behaviours such as preening, scratching, dusting, nest site avoidance, nest sanitation, migration, and other behavioural defences may drive a parasite to switch from one host species to another.

Specialist parasites with life cycles interdependent with that of hosts (e.g., in lice where the host represents the parasite's only environment), usually develop a narrow level of host specificity and only infect members of a single species which in turn determines their population structure, abundance, and prevalence [15]. In such cases, a change in host ecology can either cause a proliferation or extinction of the parasite lineage. However, Medeiros et al. [16] observed that specialists can compensate for the reduced host breadth by achieving a higher prevalence in a single host species. Generalist parasites, whose life history is not exclusively limited to one specific host, benefit from a high host diversity because they can infect multiple host species, which enables them to persist in ecosystems and potentially spread to immunologically naïve hosts, i.e., hosts that have not evolved with the parasite [13]. Host specificity is an important aspect of parasite communities and should be a key component of all wildlife disease studies because it can determine the chance of survival of a parasite in the case of a host species' extinction, the invasion potential of a parasite in new habitats such as islands, or the establishment and spread of a parasite following its introduction to a new geographical area [7].

Avian haemosporidian parasites of the genera *Plasmodium*, *Haemoproteus*, and *Leucocytozoon* are widespread vector-transmitted blood parasites that exhibit varying levels of prevalence, diversity, and host specificity at various spatiotemporal scales and across ecoregions [17]. This spatiotemporal variation could be the result of host- or parasite-mediated adaptions to the environment. Previous studies of avian haemosporidian parasites in the southern African region found varying prevalence levels, which were generally affected by the sampling effort (a low prevalence with a small sampling size and small sampling areas), while parasite lineage diversity was high as predicted [18–20]. Avian haemosporidia also exhibit varying levels of host preferences that may be driven by the host's ecological traits [19,21]. While *Haemoproteus* spp. tend to be more host specific [22], *Plasmodium* spp. have generally been found to be generalists [22,23] and *Leucocytozoon* parasites exhibit varying degrees of host specificity depending on the species and the ecological context [24]. Exceptions have also been observed where certain lineages may be restricted to a specific group of birds for all three avian haemosporidian parasite genera [25].

Despite a growing number of avian haemosporidian parasites studies, certain ecoregions and bird species remain under-sampled [26]. Few large-scale studies have been conducted on avian haemosporidia in the Afrotropical regions. In the face of global environmental change and large-scale disease emergence and spread, a substantial number of studies, especially in biodiversity hotspots, are necessary to improve our understanding of disease prevalence, diversity, and disease risk mapping. This large-scale study unravels the avian haemosporidian parasite community and host specificity at an Afrotropical biodiversity conservation region. Notwithstanding the significant role of the environment, we test the hypothesis that biodiverse regions are associated with highly diverse parasite communities. In other words, a conservation region with a high variety of potential hosts will present opportunities for a diverse parasite community. We also examined the relationship between host specificity and parasite prevalence, and test the hypothesis that host specificity, parasite prevalence, and diversity are correlated.

2. Materials and Methods

2.1. Study Area

Birds were sampled from nine sites in an Afrotropical lowveld conservation area within the greater Kruger region in South Africa. The sampling was conducted in Kruger National Park (Skukuza, Satara, Phalaborwa, Shangoni, and Shingwedzi) and the surrounding settlement areas (Acornhoek, Hazyview, Mkhuhlu, and Malelane). Sampling sizes and sampling events differed between sites and seasons where certain sites were sampled only once whilst other sites were sampled repeatedly due to ease of access and availability. More than 500 bird (both resident and migrant) species were recorded in this region [27]. This biodiversity rich region is approximately 19,600 km^2 in size; consists of three overlapping ecoregions of the world (Drakensberg montane grasslands, woodlands and forests, Southern Africa bushveld, and Zambezian and Mopane woodlands [28]; and experiences a subtropical climate (Köppen climate classification: BSh (Climate: arid; Precipitation: steppe; Temperature: hot arid)) characterised by hot wet summers (average temperature of 26.4 °C) and mild frost free dry winters (average temperature 17.8 °C). The rainy season is from September to May, with a rainfall gradient which decreases from the southern (750 mm per annum) to the northern parts of Kruger National Park (350 mm per annum).

2.2. Sampling Design and Protocol

Fieldwork was performed during the dry and wet seasons from April 2015 to November 2017. Five sampling sites were located within Kruger National Park and another four sampling sites outside the park were selected (Figure 1). Because of the nature of the park (a Big five area with security and poaching problems) and the frequent civil unrest among communities outside the park, birds were sampled opportunistically or as the sampling permits allowed and when the opportunity to visit the sites was presented. As such, certain sites were sampled only once during the three-year sampling period (Acornhoek and Malelane—2017 wet and dry) and for one single season (Hazyview—2016 wet season; Shangoni—2015 dry season; Phalaborwa—2016 dry season and 2017 wet season) whilst others were sampled for more than one year and multiple seasons were covered. Skukuza was sampled throughout the three years and during both wet and dry seasons. Satara and Shingwedzi were sampled during the wet and dry seasons of 2016 and 2017.

Live birds were sampled using birdcall lure baited mist-nets [29]. Standard morphometric measurements were taken from all captured birds (tarsus; head and culmen lengths; body mass; state of moult). Blood samples were obtained by venepuncture of the brachial vein on the right wing using a sterile 25 G needle, with blood drawn into a 75 µL micro-haematocrit capillary tube. A drop of blood was added in a vial with lysis buffer for DNA extractions and molecular detection screening. Sampled birds were released at the capture site immediately after processing.

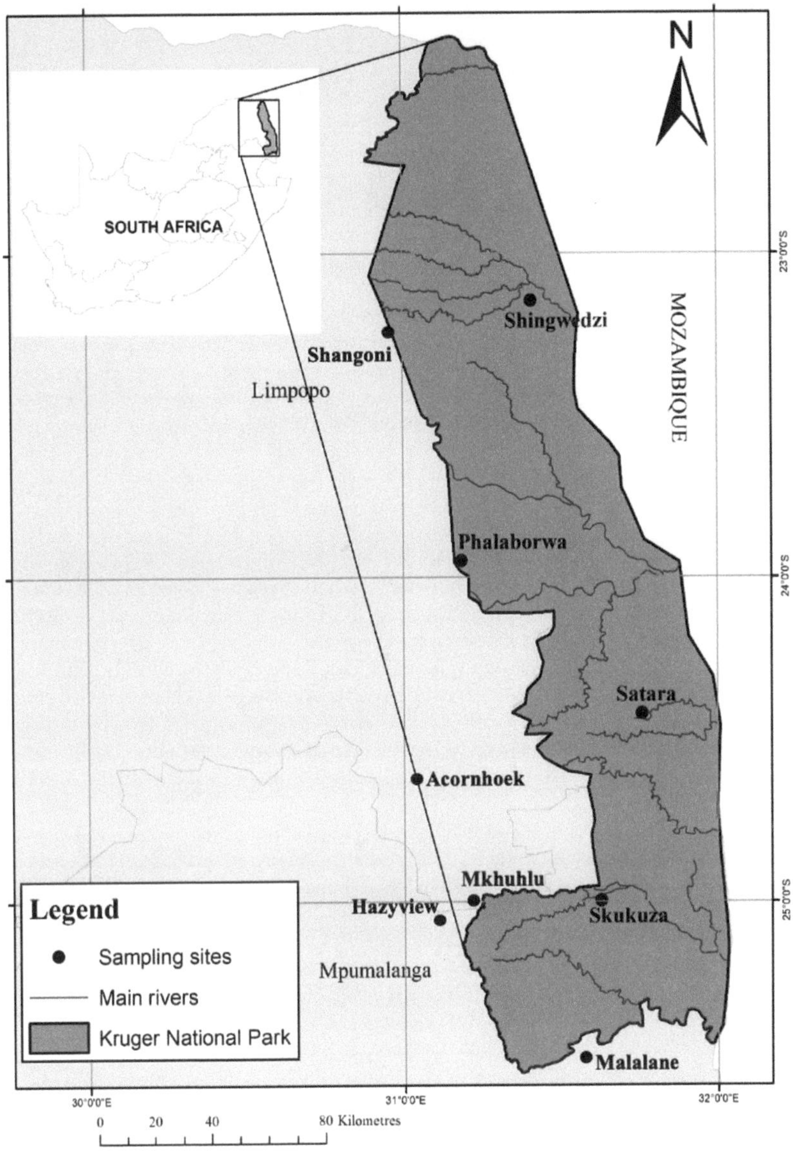

Figure 1. Geographic location of sampling sites inside and outside Kruger National Park.

2.3. Parasite Screening

To determine whether the birds harboured any avian haemosporidian parasites (genera: *Haemoproteus*, *Plasmodium*, and *Leucocytozoon*), genomic DNA was extracted from blood samples using the commercial DNeasy Blood and Tissue extraction kit (Qiagen, Valencia, CA, USA) and Invisorb Spin Blood mini kit (Stratec molecular, Berlin, Germany). Extracted DNA was quantified using the NanoDrop 2000 Spectrophotometer (Thermo Fisher Scientific, Wilmington, DE 19810 USA running the NanoDrop 2000 operating software) and then diluted to a working concentration of approximately 25 ng/µL using a TE buffer. Samples with lower DNA concentrations were not diluted. Thereafter, DNA samples were screened for haemosporidian parasites using the nested PCR protocol described by

Hellgren et al. [30]. A 479 bp fragment of the parasite's cytochrome b gene was amplified. Primer set HaemNFI and HaemNR3 were used in the first PCR to amplify the DNA of all three genera: *Plasmodium*, *Haemoproteus*, and *Leucocytozoon*. In the second PCR, we used the product of the first PCR with primer set HaemF and HaemR2 to amplify *Plasmodium* and *Haemoproteus*, and primer set HaemFL and HaemR2L for *Leucocytozoon*. A positive control (DNA template from the bird infected with *Haemoproteus*/*Plasmodium* and *Leucocytozoon*) and a negative control (distilled water) were included for every 24 samples in a 96-well plate. PCR products (1.5 μL) were checked in a 2% agarose gel stained with GelRed (Biotium, Fremont, CA, USA) using the Mupid-One electrophoresis system (Mupid Co., Ltd., Tokyo, Japan) at 50 V for 45 min and observed under UV light for bands, looking for bands of the appropriate size (479 bp). The PCR run was validated if all positive controls showed positive amplification, and the negative controls showed no amplification within the 96-well plate. Samples were run twice to confirm the results and exclude instances of false positivity or negativity. All positive PCR products were sent to Macrogen (Macrogen Inc., Amsterdam, The Netherlands) for purification and forward sequencing. Suspected new lineages were subjected to a second nested PCR protocol and sent to Macrogen for reverse sequencing to obtain the full length of the sequence.

2.4. Parasite Prevalence and Diversity

All statistical analyses were conducted using R version 4.0.1 [31] on its integrated development environment R Studio version 1.3.959 [32]. Infection prevalence was calculated as the proportion of infected individuals of each host species, determined per site as well as per parasite genus. Host species diversity (H_H) and parasite lineage diversity (H_P) of the parasite lineages infecting each host species were both calculated using Shannon's diversity index [33] as implemented in the R package vegan version 2.6 [34]. Data were tested for normality and a Pearson correlation coefficient was used to measure the linear association between the host (bird) and parasite diversity indices. A rarefaction analysis was also carried out to evaluate the completeness of our sample diversity.

2.5. Phylogenetic Analyses

Sequences obtained from Macrogen were individually checked, edited, and aligned manually using BioEdit version 7.0.5.2 [35]. The resulting sequences were individually entered in the GenBank [36] and MalAvi (accessed on 26 February 2020 [37]) databases for search and identification purposes using BLAST (Basic Local Alignment Search Tool version 2.7.1 [38]). Newly identified lineages were assigned new lineage names following the MalAvi standardized nomenclature [37] whilst sequences that were a 100% match to a lineage in the databases were assigned the corresponding MalAvi lineage name. All unique parasite lineages were subjected to phylogenetic analyses. Newly recovered parasite lineages were deposited in the GenBank database (accession numbers MW546939–94) and submitted to the MalAvi database.

A phylogenetic analysis was carried out to determine the evolutionary relationships between all the unique parasite lineages detected in this study using the Maximum Likelihood method [39]. A full phylogenetic tree was also drawn. The General Time Reversible with 5 gamma distributions (GTR + G) model was determined as the best substitution model suitable for phylogenetic reconstruction by the model function in MEGA X [40], with the lowest Bayesian Information Criteria (BIC) score. The analysis made use of 1000 bootstrap replications to generate bootstrap values.

2.6. Host Specificity

The host specificity index (S_{TD^*}) described by [7] was used to determine the host specificity of the haemosporidian lineages identified in this study. The program TaxoBiodiv2 (http://www.otago.ac.nz/parasitegroup/downloads.html, accessed on 20 April 2023) was used. The index measured the average taxonomic distinctness of all host species infected by a parasite species, weighted by the prevalence of the parasite in these different hosts, and is calculated as follows:

$$S_{TD^*} = \frac{\sum\sum_{i<j} \omega_{ij}(p_i p_j)}{\sum\sum_{i<j}(p_i p_j)}$$

where the summations are above the set ($i = 1, \ldots, s$; $j = 1, \ldots, s$, such that $i < j$ and s is the number of host species used by the parasite), ω_{ij} is the taxonomic distinctness between the host species i and j, and p_i and p_j are the prevalence of the parasite in the host species i and j, respectively [7]. In this study, the lineages that infected only one bird species were excluded from the analysis and were assigned an S_{TD^*} value of 1 as suggested by [7]. The lower values indicate parasite lineages that infect closely related hosts, while the higher values indicate parasite lineages that infect a wide range of host species.

For the host specificity index, only the lineages that infected two or more bird species were used; those detected only once were excluded since they do not provide information on the range of hosts. The construction of lineage networks for each parasite genus was performed using the medium joining network method, to test if a group of lineages or cluster of lineages were specific to certain avian families.

3. Results

3.1. Host and Parasite Diversity

A total of 1035 birds belonging to 55 species, 46 genera, 33 families, and 12 orders were sampled, of which 294 individuals (28.41%) were infected with at least one parasite genus including coinfections (Supplementary Table S1). The observed rarefaction curve indicated that our sampled bird numbers and diversity (except for Wire-tailed Swallow, *Hirundo smithii*) were indeed sufficient for the haemosporadian parasites detected (Figure 2).

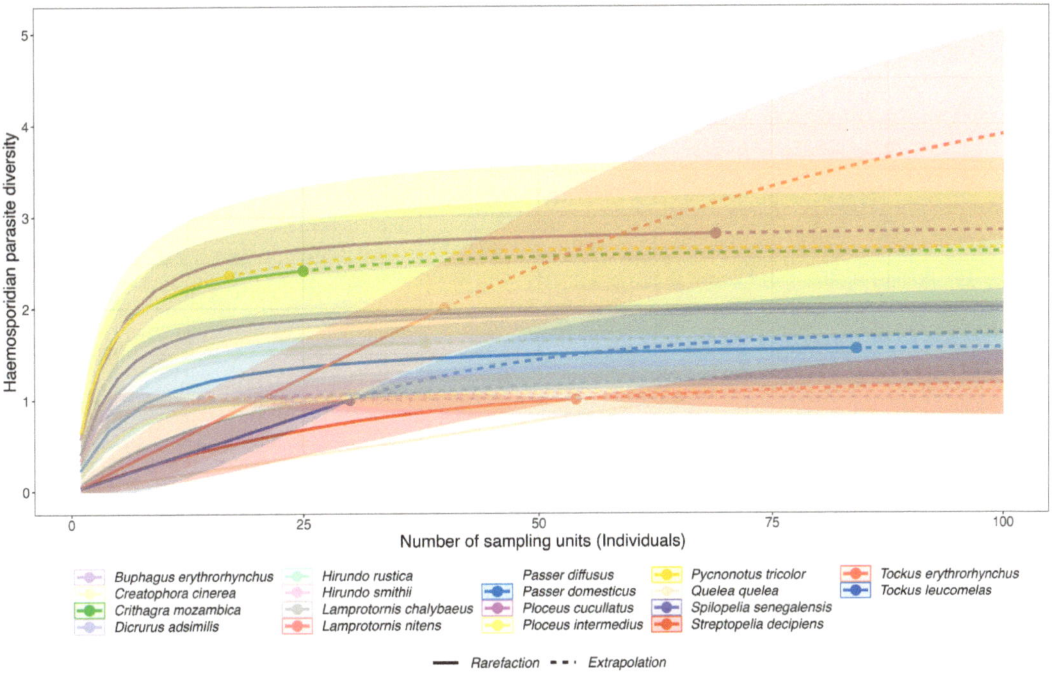

Figure 2. Rarefaction and extrapolation curves of the sampled birds against the haemosporidian parasite diversity.

The prevalence of *Haemoproteus* was 17.39% (n = 180 birds), *Leucocytozoon* was 9.28% (n = 96 birds), and *Plasmodium* infection was 4.64% (n = 48 birds, Table 1). Of the 294 infected birds, 180 were infected with *Haemoproteus* (61.22%), 48 with *Plasmodium* (16.32%), and 96 with *Leucocytozoon* (32.65%). Considering the bird species with sample sizes $\geq$ 30 birds, the highest infection prevalence for any parasite was in the Southern grey-headed Sparrows (*Passer diffusus*) and the Village Weaver (*Ploceus cucullatus*) at 65% and 55%, respectively. Whereas the lowest infection prevalence for any parasite was in the Red-billed Quelea (*Quelea quelea*) and Southern yellow-billed Hornbill (*Tockus leucomelas*) at 2% and 3%, respectively. Fifteen out of fifty-five bird species did not present any form of avian haemosporidian infection (Supplementary Table S1).

Table 1. Summary of infections per parasite genus in the lowveld region of South Africa.

	Haemoproteus	*Plasmodium*	*Leucocytozoon*	Total
No of infections detected	180	48	96	294
Prevalence (%)	17.39	4.64	9.28	28.41
Number of lineages	45	26	29	100
New identified Lineages	23	16	17	56
Existing MalAvi Lineages	22	10	12	44

One hundred distinct avian haemosporadian parasite lineages were detected for which 56 were new while 44 were already in the MalAvi database (Table 1). The most prevalent lineage was RS4 (*Leucocytozoon* sp.) and it infected the highest number of birds (n = 28 individuals, from six different bird species).

Coinfection (defined as infections with two or more different parasites, [37]), was recorded in 28 individual birds from 15 species (12 families, namely: Columbidae, n = 7; Passeridae, n = 4; Pycnonotidae, n = 4; Fringillidae, n = 3; Ploceidae, n = 3; Buphagidae, n = 1; Lybiidae, n = 1; Monarchidae, n = 1; Muscicapidae, n = 1; Paridae, n = 1; Phasianidae, n = 1; and Sturnidae, n = 1). They comprised 18 *Haemoproteus* + *Leucocytozoon* and four *Plasmodium* + *Leucocytozoon* infection combinations. The infections by *Haemoproteus* and *Plasmodium* could not be resolved. Six cases of infections by two different lineages of *Leucocytozoon* were also recorded. Another case of multiple infections by three parasite lineages was also observed in a Laughing Dove (*Spilopelia senegalensis*) which comprised one *Haemoproteus* lineage and two *Leucytozoon* lineages.

The calculated bird species diversity index among sites ranged from 1.13 to 2.51. Whereas the bird species richness varied between four and thirty-six (Table 2). In comparison, parasite lineage diversity indices per site ranged between 0 and 3.28. While parasite lineage richness was found to be between one and forty-six (Table 2). Overall, bird species diversity was positively correlated with parasite lineage diversity ($y = 2.028x - 1.805$, $r = 0.866$, $F = 21.059$, $p = 0.0025$, Figure 3).

Table 2. A comparison of bird and parasite lineage species richness (S_H and S_P) and diversity indices (H_H an H_P) calculated for each sampling site (inside vs. outside Kruger National Park).

			Host Species		Parasite Lineage	
Location	Sampling Sites	Coordinates	S_H	H_H	S_P	H_P
Outside Kruger National Park	Acornhoek	31.041156° −24.587340°	10	1.84	13	2.46
	Hazyview	31.185619° −25.032365°	4	1.13	1	0
	Malelane	31.574146° −25.467964°	11	1.73	7	1.91
	Mkhuhlu	31.241542° −24.995130°	15	2.13	22	2.72

Table 2. *Cont.*

Location	Sampling Sites	Coordinates	Host Species		Parasite Lineage	
			S_H	H_H	S_P	H_P
Inside Kruger National Park	Phalaborwa	31.169120° −23.937940°	16	2.51	14	2.56
	Satara	31.774039° −24.397732°	19	2.08	21	2.77
	Shangoni	30.975002° −23.239999°	7	1.89	4	1.39
	Shingwedzi	31.425900° −23.113545°	11	1.82	14	2.34
	Skukuza	31.603911° −24.996356°	36	2.46	46	3.28

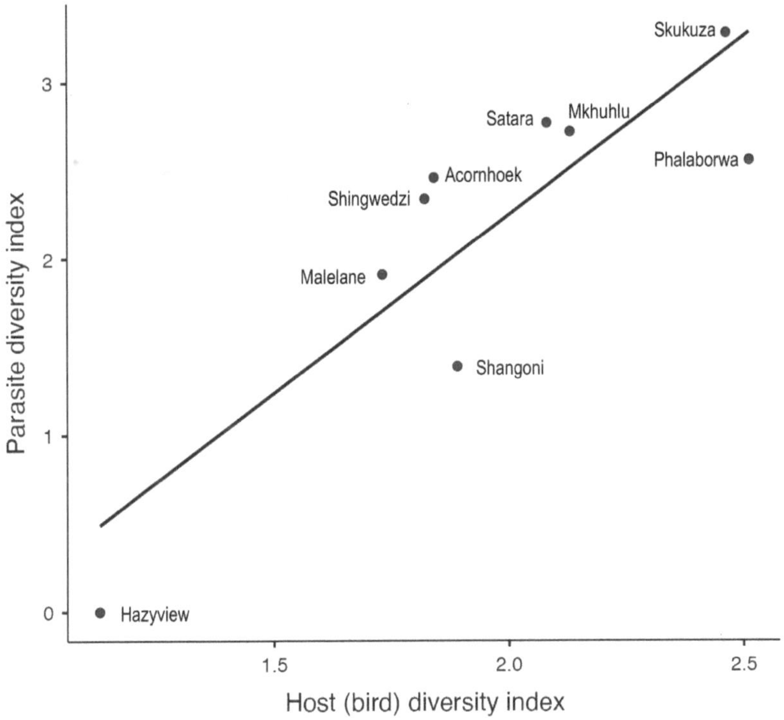

Figure 3. Relationship between bird diversity and parasite lineage diversity indices across the nine sampling sites inside and adjacent to Kruger National Park. Fitted linear regression: $y = 2.028x - 1.805$; $r = 0.866$; $F = 21.059$; $p = 0.0025$.

Four *Haemoproteus* lineages (AFR041, AFR076, CRECRI01 and SPISEN02) infected the most individuals whilst three *Leucocytozoon* lineages (REB7, RS4, and SPISEN06) and one *Plasmodium* lineage (TOCERY01) were also common (Table S1). The birds infected with the highest number of parasite lineages were the Village Weavers (19 lineages), Greater blue-eared Starlings (*Lamprotornis chalybaeus*; 12 lineages), and Southern grey-headed Sparrows (12 lineages). All three are resident species.

3.2. Phylogenetic Relationships

The individual parasite genus phylogenetic trees are provided separately (in sub-trees) for better visibility of the relationship between the lineages (Figures 4–6). The *Plasmodium* phylogenetic tree formed five clusters, with the largest cluster comprising eight lineages (Figure 4). The smallest cluster contained only two parasite lineages. The lineage PELSEP04, a novel parasite lineage detected from a Crested Francolin (*Peliperdix sephaena*), was unique and differed by 24 bp from its most closely related lineage in the MalAvi and Genbank databases, with a 95% similarity to the lineage TURPEL05 (GenBank accession number: MG018674.1).

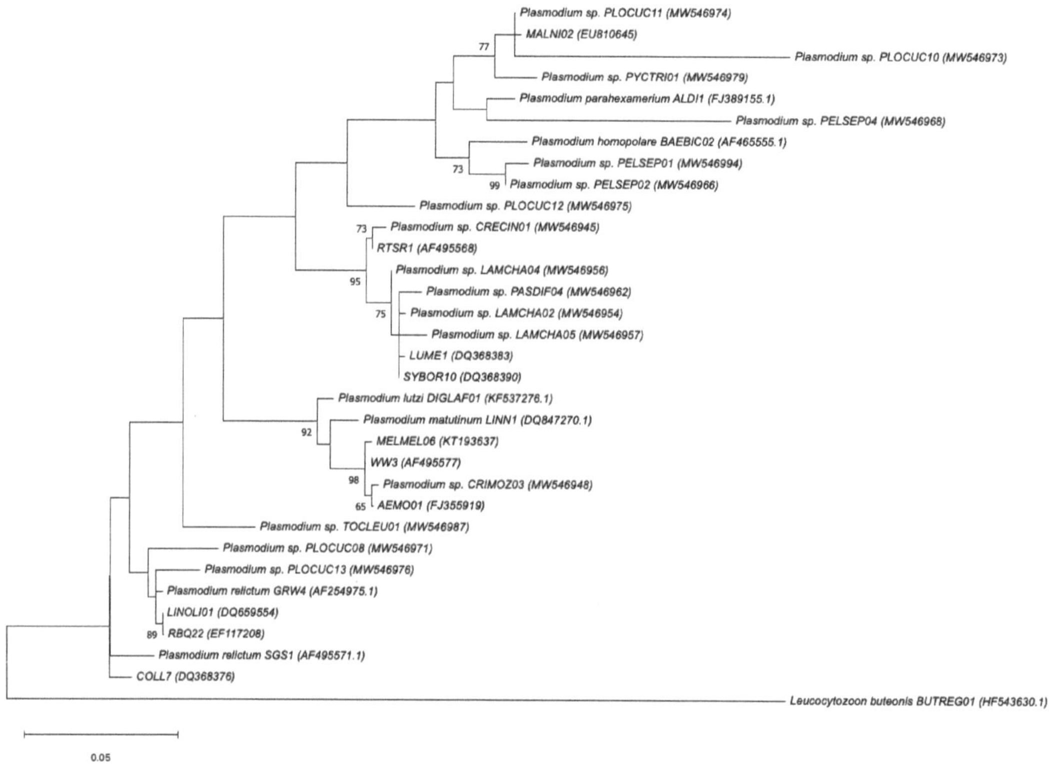

Figure 4. Phylogenetic relationship between *Plasmodium* parasite lineages (479 bp mitochondrial cytochrome b) detected inside and adjacent to Kruger National Park, as determined by the Maximum Likelihood (ML) method. ML bootstrap values >50% are indicated next to the branch nodes. The closest described species match from the NCBI BLAST search, for each cluster, is included in the tree. GenBank accession numbers are included in brackets.

The *Leucocytozoon* tree identified three clusters, including the largest cluster (with 22 parasite lineages) and two small clusters with two and three lineages (Figure 5). All 22 lineages in the largest cluster showed between 96 and 99% similarity to *Leucocytozoon gentili*. Similar to lineage PELSEP04 (*Plasmodium* sp.) described above, lineage PELSEP03 (*Leucocytozoon* sp.) was unique, and it was also found in the same individual. PELSEP03 was 95% similar to *Leucocytozoon schoutedeni*, which was previously recovered and described in chickens (lineages GALLUS06 and GALLUS07; accession numbers DQ676823 and DQ676824, respectively). Both lineages PELSEP03 and PELSEP04 were newly identified lineages in this study and were a coinfection case in the same bird and the first record in this species. Perhaps this is indicative of some level of specialisation of these two lineages.

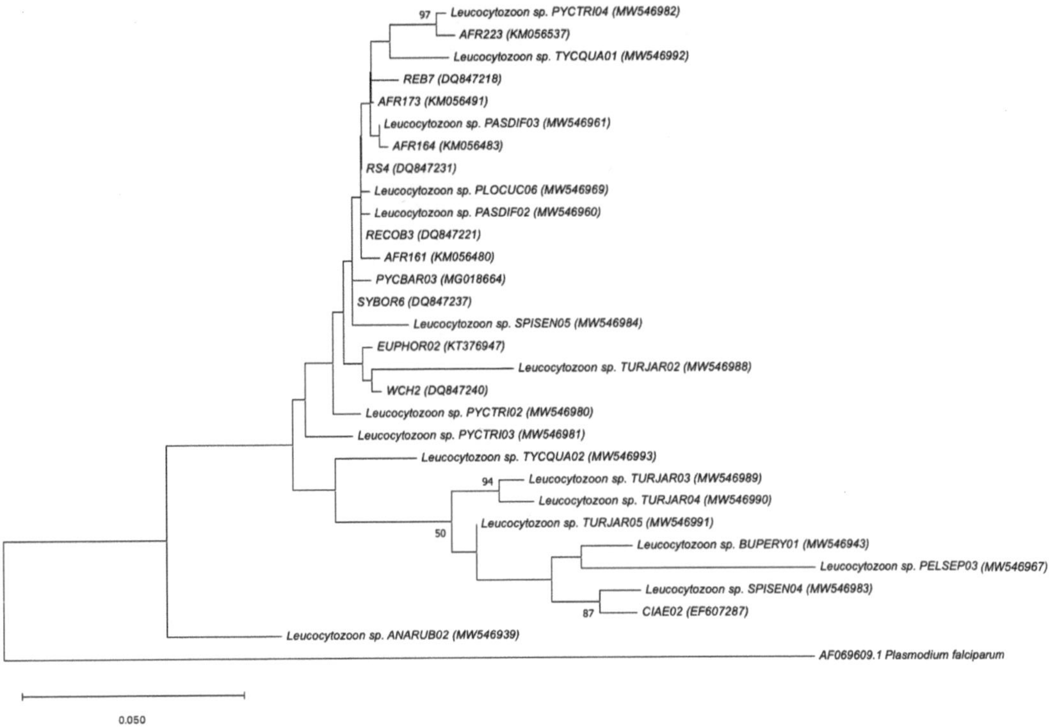

Figure 5. Phylogenetic relationship of *Leucocytozoon* parasite lineages (479 bp mitochondrial cytochrome b lineages) detected inside and adjacent to Kruger National Park, as determined by Maximum Likelihood (ML) method. ML bootstrap values >50% are indicated next to the branch nodes. GenBank accession numbers are included in brackets.

The *Haemoproteus* tree identified seven clusters with the largest clusters comprising seven parasite lineages each (Figure 6). Twelve lineages did not cluster, of which six of those new lineages were detected for the first time in this study. Six previously reported lineages did not fall within the identified clusters and stood on their own. The *Haemoproteus* lineages appeared to be more diverse and less related than the *Plasmodium* and *Leucocytozoon* lineages. Most of the newly identified *Haemoproteus* lineages fell within the identified clusters. The lineage BUTVER01 (*Haemoproteus* sp.) was 96% similar to the lineage CATAUR01 in the MalAvi and GenBank databases (GenBank accession number MF953291), was described as *Haemoproteus catharti*, and was found in a Water thick-knee (*Burhinus vermiculatus*).

A comparison of the parasite lineages with closely described species in the Genbank and MalAvi databases revealed varying levels of diversity. The largest *Plasmodium* cluster grouped parasite lineages were unknown (unclassified). The second largest cluster aligned with *Plasmodium relictum* (96% to 100% similarity). *Plasmodium parahexamerium* and *Plasmodium homopolare* had similarity percentages ranging between 95% and 97% with the lineages PELSEP01, PELSEP02, and PELSEP04. *Plasmodium lutzi* and *Plasmodium matutinum* aligned with CRIMOZ03 with 97% and 98% similarity, respectively.

The *Leucocytozoon* lineages recorded in this study mainly represented two known *Leucocytozoon* species: *Leucocytozoon gentili* (22 lineages) and *Leucocytozoon californicus* (MalAvi ID FASPA02; three lineages). The *Haemoproteus* clusters were common and the most diverse represented the following species *Haemoproteus homobelopolskyi, Haemoproteus homominutus, Haemoproteus homopalloris, Haemoproteus homopicae, Haemoproteus lanii, Haemoproteus pallidus,*

Haemoproteus paranucleophilus, *Haemoproteus pastoris*, *Haemoproteus sacharovi*, *Haemoproteus sanguinis*, and *Parahaemoproteus passeris*.

Figure 6. Phylogenetic relationship of *Haemoproteus* parasite lineages (479 bp mitochondrial cytochrome b lineages) detected inside and adjacent to Kruger National Park, as determined by the Maximum Likelihood (ML) method. ML bootstrap values (>50%) from 1000 replications are indicated next to the branch nodes. The closest described species match from the NCBI BLAST search for each cluster is shown. GenBank accession numbers are included in brackets.

3.3. Host Specificity

The eight *Haemoproteus*, six *Plasmodium*, and four *Leucocytozoon* lineages infected at least two host species. The calculated S_{TD^*} values for all three parasite genera ranged between one and four and indicated the marked variation in host specificity. The parasite lineage that infected the highest number ($n = 6$) of bird species was RS4 (*Leucocytozoon* sp.; Table 3). For *Plasmodium*, three out of six lineages (RTSR1, SYBOR10, and TOCERY01) had an S_{TD^*} value larger than three, suggesting that there were generalists. The lineage RBQ22 from *Plasmodium*, with an S_{TD^*} value of three was also identified as a generalist, which was recorded in the Lesser striped Swallow (*Cecropis abyssinica*) and the Village Weaver (*Ploceus cucullatus*) who belong to the superfamilies Sylvioidea and Passeroidea, respectively. For *Haemoproteus*, most of the lineages were recorded as specialists. It was, however, noted that the lineages SPISEN02 and AFR173 (*Haemoproteus* sp.) had the highest S_{TD^*} values of four and three, respectively, and infected two unrelated species, indicative of a generalist lineage. For *Leucocytozoon*, two of the lineages (PASDIF03 and LAMCHA04) were generalists and the other two (REB7 and RS4) were specialists. In general, when considering only the lineages that infected two or more hosts in this study, *Plasmodium* and *Leucocytozoon* recorded the average S_{TD^*} values of 2.44 and 2.48, respectively, and were identified as generalists, whilst *Haemoproteus* had an average index of 1.76 indicating that the lineages found were specialists.

Table 3. Number of hosts infected by each parasite lineage and their corresponding specificity index (S_{TD^*}). Only lineages that infected more than one host are presented.

Genus	Lineage	Host Infected	S_{TD^*}	Host Specificity
Haemoproteus	AFR041	2	1	Specialist
	AFR076	2	1	Specialist
	AFR084	2	2	Specialist
	PAMEL01	3	1.98	Specialist
	SPISEN02	2	4	Generalist
	VILWE2	4	1.32	Specialist
	VIMWE1	3	1	Specialist
	AFR173	2	3	Generalist
Leucocytozoon	PASDIF03	2	1	Specialist
	REB7	4	3	Generalist
	RS4	6	2.90	Generalist
	LAMCHA04	2	1	Specialist
Plasmodium	LAMCHA05	2	1	Specialist
	MALNI02	2	1	Specialist
	RBQ22	2	3	Generalist
	RTSR1	3	3.98	Generalist
	SYBOR10	4	3.73	Generalist
	TOCERY01	3	3.37	Generalist

Haplotypes were specific to avian families.

4. Discussion

Although studies on parasite–host interactions have greatly improved our understanding of the co-evolutionary relationship between a parasite and its host, a lot remains to be deciphered. Variations in the host specificity of parasite species (i.e., the diversity of the host species a parasite infects, ranging from specialist to generalist parasites) have shown to markedly differ from one region to another, driven by both the host's traits and environmental factors (i.e., biomes and geographical barriers; [41]). In this study, we examined parasite prevalence, diversity, and host specificity in a community of avian haemosporidian parasites and explained these in relation to avian host diversity and sampling localities. Our findings, even though some samples were collected seven years ago, revealed a total avian haemosporidian infection prevalence of 28.41% which varied between parasite genera. There was also a relatively high degree of diversity in bird host species and parasite lineages, as well as a varying degree of host specificity among parasite lineages. Previous studies in other Afrotropical regions reported a higher avian haemosporidian infection prevalence than observed in this study, and a high level of parasite lineage diversity [20,42]. The authors in Illera et al. [43] also suggested that host richness might explain the variations in parasite prevalence and richness. In their study, *Plasmodium* prevalence and richness showed a negative correlation with host richness, whereas *Haemoproteus* and *Leucocytozoon* were positively correlated with host richness. The observed prevalence in this study could be explained by the dilution effect hypothesis [44–47]. This hypothesis posits that diverse ecosystems will limit disease transmission and spread, thus leading to a low disease prevalence [44]. In essence, it is in the interest of global public health to conserve biodiversity since rich and diverse communities will dilute the effect of parasites by lowering the risk of disease and reducing the rates of pathogen transmission. Hence, Civitello et al. [48] stressed that human-induced biodiversity loss would increase wildlife disease prevalence. Conversely, it is worth noting that a positive relationship (an amplification effect) between pathogen prevalence and biodiversity may occur [49]. The authors in Roiz et al. [50] for example, recorded a higher prevalence of the Usutu virus in areas with richer avian communities. This study did not exhaust all other elements, specifically assessing factors driving dilution, but perhaps it should be a topic to explore in another study.

Despite the relatively low prevalence recorded in this study compared to other Afrotropical regions, a high parasite lineage diversity was recorded here, corroborating previous findings and predictions from the region and globally [23,51–53]. The authors in Chaisi et al. [51] recorded an overall parasite prevalence of 68.82% in the Afrotropical terrestrial birds from 93 samples collected in South Africa (n = 76) and West Africa (N = 17), whilst Lutz et al. [20] recorded 79.1% from 532 birds sampled in Malawi. They also recorded an exceptionally high parasite diversity with 248 parasite cytochrome *b* lineages identified from 152 host species. The authors in Outlaw et al. [52], on the other hand, demonstrated that avian haemosporidian parasites exhibit a similar pattern of diversity to their hosts and suggested that parasites should be the most diverse in regions with the greatest proportions of endemic host species. Our study area was conducted in an Important Bird and Biodiversity Area (IBA), and, with the exception of four migrant bird species (Violet-backed Starling *Cinnyricinclus leucogaster*, Barn Swallow *Hirundo rustica*, Red-backed Shrike *Lanius collurio*, and the African Paradise Flycatcher *Terpsiphone viridis*), all other birds sampled during this study are endemic to the region, suggesting a link between host endemicity and the high diversity of the parasite lineages recorded (100 unique parasite lineages from 294 infected individuals). This assertion is further confirmed by the observed positive relationship between the bird and parasite diversity indices and confirmed by Wardjomto et al. [54]. In line with these findings, where parasite diversity is expected to increase with improved sampling, perhaps more bird samples and species diversity would unravel an elevated infection prevalence and diversity. Our current sample of the diversity (55 out of 490 species) represents approximately 11% of all bird species in this region. Nevertheless, the contribution of this study is significant since new avian haemosporidian lineages were found in 12 resident bird species, from the 55 bird species that were sampled.

The parasite diversity observed in this study can also be attributed to the diversity of the habitats, which sustain large populations of host species intertwined in a multitude of ecological and co-evolutionary processes with the vectors—we will further explore this in other studies. In essence, the ubiquitous nature of the *Plasmodium* vector (generally *Culex* mosquito, [24]) and the endemicity of human malaria in the region suggest that the habitat and environmental conditions are suitable for vector proliferation and disease prevalence. The vectors of the *Haemoproteus* and *Leucocytozoon* parasites, which require a semi-moist to arid habitat [24], may also find a suitable habitat in this Afrotropical region.

Most bird species sampled were infected by a single parasite lineage, indicating a high degree of host specificity in the area. Host specificity, in this case, could explain the low prevalence. The host-specific parasite lineage may infect a single or closely related species and thus affect their prevalence in the area, especially if the host distribution and numbers are limited. The longitudinal and latitudinal variation in the host species in the area is expected to limit the host distribution and therefore, affect parasite prevalence [55]. As observed here, sites inside Kruger National Park had a marginally higher infection (prevalence = 30%) than those outside (prevalence = 23%) the park, supporting the assertion that avian haemosporidian parasites are more prevalent in undisturbed areas than disturbed areas [11]. Although, urban and arid environments may hinder the development of competent vectors [11], thus keeping their prevalence relatively low in the region.

The host specificity index used in this study measured the average taxonomic distinctiveness weighted by the prevalence of the parasite in different hosts [7] and considered the haemoparasite species that infected two or more hosts for comparison purposes. This index is such that the lineages infected by one parasite only are assigned a value of zero or one or omitted altogether [7] with their host specificity classified as undetermined. The value of S_{TD^*} increases as the taxonomic distinctiveness between the high-prevalence hosts increases, although the effect on S_{TD^*} is greater with a more drastic change in the taxonomic distance than with changes in prevalence. The average observed host specificity index (S_{TD^*}) values of 2.44 and 2.48 for *Plasmodium* and *Leucocytozoon*, respectively, suggest that these parasite genera were infecting birds that are distantly related (from different families or different orders) whilst *Haemoproteus* was infecting more closely related bird hosts (generally constrained to the family level). This finding is in line with those of [56] in Madagascar where the *Haemoproteus* lineages were mainly host-specific and the *Plasmodium* and *Leucocytozoon* lineages were generalist. The proximity of Madagascar to the mainland of Southern Africa could explain this similarity. Although the *Leucocytozoon* lineages are generally host specific [42], transitions from generalist to specialist and vice versa are common [3,57]. These transitions are suspected to be an evolutionary strategy driven by competition, climate change, and large-scale ecological perturbation [57,58]. By probing only the *Haemoproteus* Lineage SPISEN02 identified as a generalist (with the highest host specificity index of S_{TD^*} = 4), it was observed that it was present in two distantly related bird species (Red-headed Weaver: *Anaplectes rubriceps* and Laughing Dove) with equal prevalence value of 10% thus driving the value of S_{TD^*} up. Until now, this lineage had only been detected in Laughing Doves (evidenced by the MalAvi database; [37]). It is therefore possible that this lineage, although identified as a generalist, could be a host specific lineage. However, more data are needed to confirm this assertion. The *Leucocytozoon* lineages AFR173, RB7, and RS4 were found in birds of different orders and classes (i.e., highest taxonomic distance), confirming that these were generalist infections.

For the existing parasite lineages with undetermined host specificity in this study, the MalAvi database shed some light on their host specificity status. The lineage BUL2 (*Haemoproteus sanguinis*) was recorded infecting closely related bird species in the MalAvi database (Willow Warbler *Phylloscopus trochilus*, Dark-capped Bulbul *Pycnonotus tricolor*, White-spectacled Bulbul *Pycnonotus xanthopygos*, Malagasy Bulbul *Hypsipetes madagascariensis*, and African red-eyed Bulbul *Pycnonotus nigricans*) mainly from the African continent and predominantly infecting the Dark-capped Bulbul (*Pycnonotus tricholor*) observed in this study. The lineage BUL2 could therefore be classified as host-specific, infecting closely related species [37]. The lineages CRECRI01, CRECIN01, and CRECIN02 are also host-specific infecting only the Wattled Starling (*Creatophora cinerea*). In essence, except for SPISEN01 (infecting *Passer domesticus* and *Streptopelia senegalensis*), nearly all the *Haemoproteus* lineages recorded in this study were host-specific. Similarly, except for the parasite lineages recorded for the first time in this species, the existing *Plasmodium* lineages (AEMO01, COLL7, LINOLI01, MELMEL06, MALNI02, RBQ22, RTSR1, and SYBOR10) appeared to be generalists. Furthermore, except for the lineages recorded for the first time in this study, all existing MalAvi *Leucocytozoon* lineages recorded in this study (AFR161, AFR164, AFR173, EUPHOR02, PASDIF03, REB7, RECOB3, RS4, SYBOR06, and WCH2) also appeared to be generalists.

The relationship between the parasite lineages and infection patterns confirmed the findings of the host specificity pattern of the lineages recovered in this study, although exceptions were observed. The parasite lineages from the largest cluster of the *Haemoproteus* genus were found to be infecting other bird species in the MalAvi database as well, suggesting that parasite lineages from this cluster are generalists. A high level of parasite lineage endemicity was observed among the parasite lineages recorded in this study. This was evident in the high number of new parasite lineages recorded and the lack of described, closely related, parasite species for a large number of the lineages recorded in this study (26 lineages in total). Six parasite lineages (COLL7, LINOLI01, PLOCUC08, PLOCUC13, RBQ22, and TOCLEU01) were closely related to *Plasmodium relictum*, the causal agent of avian malaria responsible for the mortality and extinction of several Hawaiian bird species [59]. It is, however, noted that the level of pathology may differ for closely related parasite lineages and strains in different bird species [37]. Further studies may be necessary to provide clarity on the extent of the effect of closely related parasite strains on avian hosts.

5. Conclusions

This study is the first large-scale study to molecularly describe avian haemosporidian parasites in the Afrotropical conservation area around Kruger National Park in South Africa. A low parasite prevalence was observed in this study, but a high parasite diversity and a large number of new parasite lineages were identified which will contribute to enriching the existing avian malaria and associated haemosporidian parasites database (MalAvi). The *Haemoproteus* lineages were generally specialist whilst the *Plasmodium* and *Leucocytozoon* lineages were generalist, resulting in a marked phylogenetic structure. The observed positive relationship between avian host diversity and parasite diversity is indicative of an amplification effect. These findings provide the opportunity to test new hypotheses to improve our understanding of host–parasite co-evolution, the drivers of parasite infection, the prevalence and diversity of parasites in a fairly natural setting, as well as the factors that lead to host specialisation and generalisation.

Supplementary Materials: The following supporting information can be downloaded at: https://www.mdpi.com/article/10.3390/ani14192906/s1, Table S1: summary of all birds sampled, as well as the prevalence and corresponding parasite lineages recorded in each infected bird host. Title headings: n = sample size; H = *Haemoproteus* spp.; P = *Plasmodium* spp.; and L = *Leucocytozoon*. The bold lineages indicate new lineages. The numbers in the grey column indicate the sample sizes of the bird(s) infected by a lineage for that species.

Author Contributions: M.N., M.B.W. and T.C.N. designed the study; M.N., T.P. and M.B.W. collected field data; M.B.W. and T.C.N. conducted all molecular analyses; All authors contributed to the writing of this manuscript. All authors have read and agreed to the published version of the manuscript.

Funding: Funding for this research came from the National Research Foundation (Grant number: IFR190125410940), the Foundational Biodiversity Information Programme (Grant number: FBIP190314423449), and the University of the Free State.

Institutional Review Board Statement: This study was approved by the South African National Park (Research Permit No: NDLM1262) and the University of the Witwatersrand Animal Ethics Screening Committee (Clearance Certificate No: 2015/02/B). Other ethical clearance was obtained for the overarching project registered by M Ndlovu at the University of the Free State (UFS-AED2017/0004). Permission to conduct this research in terms of section 20 of the Animal Disease Act 1984 (Act No 35 of 1984) was obtained from the Department of Agriculture, Forestry, and Fisheries of South Africa (Ref number 12/11/1/4). All bird capture and ringing were conducted in accordance with the SAFRING (South African Ringing Scheme) bird ringing manual under the supervision of M Ndlovu.

Informed Consent Statement: Not applicable.

Data Availability Statement: Data for this study are freely available from the South African Foundational Biodiversity Information Programme (FBIP) repository and are deposited at the following link: https://figshare.com/articles/dataset/Data_01_12_2020_xlsx/13317293 (assessed on the 1 December

2020). Accession numbers of the new parasite lineages detected are available in the GenBank and MalAvi databases.

Acknowledgments: We thank Alfredo Colina, Miranda Thomas, Damaris Chenoweth, Jackson Seminara, and several other undergraduate students from the Organisation for Tropical Studies for their help with the fieldwork. Antón Perez-Rodríguez assisted with the lab work and data analysis. The South African National Parks (SANParks) and the Mpumalanga local Municipalities kindly granted permission for the birds to be captured in Kruger National Park and the sites adjacent to the park, respectively.

Conflicts of Interest: The authors declare no conflicts of interest. The funders had no role in the design of this study; in the collection, analyses, or interpretation of the data; in the writing of this manuscript; or in the decision to publish the results.

References

1. Vourc'h, G.; Plantard, O.; Morand, S. How Does Biodiversity Influence the Ecology of Infectious Disease? In *New Frontiers of Molecular Epidemiology of Infectious Diseases*; Springer: Dordrecht, The Netherlands, 2011; pp. 291–309. [CrossRef]
2. Licht, H.H.D.F. Does pathogen plasticity facilitate host shifts? *PLoS Pathog.* **2018**, *14*, e1006961. [CrossRef]
3. Huang, S.; Drake, J.M.; Gittleman, J.L.; Altizer, S. Parasite diversity declines with host evolutionary distinctiveness: A global analysis of carnivores. *Evolution* **2015**, *69*, 621–630. [CrossRef] [PubMed]
4. Chakraborty, D.; Reddy, M.; Tiwari, S.; Umapathy, G. Land Use Change Increases Wildlife Parasite Diversity in Anamalai Hills, Western Ghats, India. *Sci. Rep.* **2019**, *9*, 11975. [CrossRef]
5. Budria, A.; Candolin, U. How does human-induced environmental change influence host-parasite interactions? *Parasitology* **2014**, *141*, 462–474. [CrossRef]
6. Wells, K.; Clark, N.J. Host Specificity in Variable Environments. *Trends Parasitol.* **2019**, *35*, 452–465. [CrossRef]
7. Poulin, R.; Mouillot, D. Combining phylogenetic and ecological information into a new index of host specificity. *J. Parasitol.* **2005**, *91*, 511–514. [CrossRef] [PubMed]
8. Padilla, D.P.; Illera, J.C.; Gonzalez-Quevedo, C.; Villalba, M.; Richardson, D.S. Factors affecting the distribution of haemosporidian parasites within an oceanic island. *Int. J. Parasitol.* **2017**, *47*, 225–235. [CrossRef]
9. Galen, S.C.; Ray, S.; Henry, M.; Weckstein, J.D. Parasite-associated mortality in birds: The roles of specialist parasites and host evolutionary distance. *Biol. Lett.* **2022**, *18*, 20210575. [CrossRef]
10. Ventim, R.; Morais, J.; Pardal, S.; Mendes, L.; Ramos, J.A.; Pérez-Tris, J. Host-parasite associations and host-specificity in haemoparasites of reed bed passerines. *Parasitology* **2012**, *139*, 310–316. [CrossRef]
11. Sehgal, R.N.M. Manifold habitat effects on the prevalence and diversity of avian blood parasites. *Int. J. Parasitol. Parasites Wildl.* **2015**, *4*, 421–430. [CrossRef]
12. Santiago-Alarcon, D.; Havelka, P.; Pineda, E.; Segelbacher, G.; Schaefer, H.M. Urban forests as hubs for novel zoonosis: Blood meal analysis, seasonal variation in Culicoides (Diptera: Ceratopogonidae) vectors, and avian haemosporidians. *Parasitology* **2013**, *140*, 1799–1810. [CrossRef] [PubMed]
13. Garcia-Longoria, L.; Marzal, A.; de Lope, F.; Garamszegi, L. Host-parasite interaction explains variation in the prevalence of avian haemosporidians at the community level. *PLoS ONE* **2019**, *14*, e0205624. [CrossRef] [PubMed]
14. Bush, S.E.; Clayton, D.H. Anti-parasite behaviour of birds. *Philos. Trans. R. Soc. B Biol. Sci.* **2018**, *373*, 20170196. [CrossRef]
15. Martinů, J.; Hypša, V.; Štefka, J. Host specificity driving genetic structure and diversity in ectoparasite populations: Coevolutionary patterns in Apodemus mice and their lice. *Ecol. Evol.* **2018**, *8*, 10008–10022. [CrossRef]
16. Medeiros, M.C.I.; Ellis, V.A.; Ricklefs, R.E. Specialized avian Haemosporida trade reduced host breadth for increased prevalence. *J. Evol. Biol.* **2014**, *27*, 2520–2528. [CrossRef] [PubMed]
17. Svensson-Coelho, M.; Blake, J.G.; Loiselle, B.A.; Penrose, A.S.; Parker, P.G.; Ricklefs, R.E. Diversity, Prevalence, and Host Specificity of Avian *Plasmodium* and *Haemoproteus* in a Western Amazon Assemblage—Diversity, Prevalence, and Host Specificity of Avian *Plasmodium* and *Haemoproteus* in a Western Amazon Assemblage. *Ornithol. Monogr.* **2013**, *76*, 1–47. [CrossRef]
18. Earle, R.A.; Bennett, G.F.; Du, T.H.; De, S.D.H.; Herholdt, J.J. Regional and seasonal distribution of avian blood parasites from northern South Africa. *S. Afr. J. Wildl. Res.* **1991**, *21*, 47–53.
19. Okanga, S.; Cumming, G.S.; Hockey, P.A. Avian malaria prevalence and mosquito abundance in the Western Cape, South Africa. *Malar. J.* **2013**, *12*, 370. [CrossRef]
20. Lutz, H.L.; Hochachka, W.M.; Engel, J.I.; Bell, J.A.; Tkach, V.V.; Bates, J.M.; Hackett, S.J.; Weckstein, J.D. Parasite Prevalence Corresponds to Host Life History in a Diverse Assemblage of Afrotropical Birds and Haemosporidian Parasites. *PLoS ONE* **2015**, *10*, e0121254. [CrossRef]
21. Gupta, P.; Vishnudas, C.K.; Robin, V.V.; Dharmarajan, G. Host phylogeny matters: Examining sources of variation in infection risk by blood parasites across a tropical montane bird community in India. *Parasites Vectors* **2020**, *13*, 536. [CrossRef]
22. Ferraguti, M.; Martínez-de la Puente, J.; Muñoz, J.; Roiz, D.; Ruiz, S.; Soriguer, R.; Figuerola, J. Avian *Plasmodium* in *Culex* and *Ochlerotatus* Mosquitoes from Southern Spain: Effects of Season and Host-Feeding Source on Parasite Dynamics. *PLoS ONE* **2013**, *8*, e66237. [CrossRef] [PubMed]

23. Clark, N.J.; Clegg, S.M.; Lima, M.R. A review of global diversity in avian haemosporidians (*Plasmodium* and *Haemoproteus*: Haemosporida): New insights from molecular data. *Int. J. Parasitol.* **2014**, *44*, 329–338. [CrossRef] [PubMed]
24. Valkiunas, G.; Iezhova, T.A.; Krizanauskiene, A.; Palinauskas, V.; Sehgal, R.N.M.; Bensch, S. A comparative analysis of microscopy and PCR-based detection methods for blood parasites. *J. Parasitol.* **2008**, *94*, 1395–1401. [CrossRef] [PubMed]
25. Moens, M.; Pérez-Tris, J. Discovering potential sources of emerging pathogens: South America is a reservoir of generalist avian blood parasites. *Int. J. Parasitol.* **2015**, *46*, 41–49. [CrossRef]
26. Harvey, J.A.; Voelker, G. Host associations and climate influence avian haemosporidian distributions in Benin. *Int. J. Parasitol.* **2019**, *49*, 27–36. [CrossRef]
27. Tarboton, W.; Ryan, P.G. *Guide to Birds of the Kruger National Park*; Struik Nature, Penguin Random House: Cape Town, South Africa, 2016.
28. Olson, D.M.; Dinerstein, E.; Wikramanayake, E.D.; Burgess, N.D.; Powell, G.V.N.; Underwood, E.C.; D'amico, J.A.; Itoua, I.; Strand, H.E.; Morrison, J.C.; et al. Terrestrial Ecoregions of the World: A New Map of Life on Earth: A new global map of terrestrial ecoregions provides an innovative tool for conserving biodiversity. *BioScience* **2001**, *51*, 933–938. [CrossRef]
29. Ndlovu, M. Birdcall lures improve passerine mist-net captures at a sub-tropical African savanna. *PLoS ONE* **2018**, *13*, e0199595. [CrossRef]
30. Hellgren, O.; Waldenström, J.; Bensch, S. A new PCR assay for simultaneous studies of *Leucocytozoon*, *Plasmodium*, and *Haemoproteus* from avian blood. *J. Parasitol.* **2004**, *90*, 797–802. [CrossRef]
31. R Core Team. *R: A Language and Environment for Statistical Computing*; R Foundation for Statistical Computing: Vienna, Austria, 2021.
32. RStudio Team. *RStudio: Integrated Development for R*; RStudio Team: Boston, MA, USA, 2020.
33. Shannon, C.E. A Mathematical Theory of Communication. *Bell Syst. Tech. J.* **1948**, *27*, 379–423. [CrossRef]
34. Oksanen, J.; Simpson, G.L.; Blanchet, F.G.; Kindt, R.; Legendre, P.; Minchin, P.R.; O'Hara, R.B.; Solymos, P.; Stevens, M.H.H.; Szoecs, E.; et al. vegan: Community Ecology Package. 2022. Available online: https://CRAN.R-project.org/package=vegan (accessed on 21 April 2023).
35. Hall, T.A. Bioedit: A User-Friendly Biological Sequence Alignment Editor and Analysis Program for Windows 95/98/nt. Nucleic Acids Symposium Series. 1 January 1999. Available online: https://www.scinapse.io/papers/1483247593 (accessed on 21 April 2023).
36. Sayers, E.W.; Cavanaugh, M.; Clark, K.; Ostell, J.; Pruitt, K.D.; Karsch-Mizrachi, I. GenBank. *Nucleic Acids Res.* **2020**, *48*, D84–D86. [CrossRef]
37. Bensch, S.; Stjernman, M.; Hasselquist, D.; Ostman, O.; Hansson, B.; Westerdahl, H.; Pinheiro, R.T. Host specificity in avian blood parasites: A study of *Plasmodium* and *Haemoproteus* mitochondrial DNA amplified from birds. *Proc. Biol. Sci.* **2000**, *267*, 1583–1589. [CrossRef]
38. Altschul, S.F.; Gish, W.; Miller, W.; Myers, E.W.; Lipman, D.J. Basic local alignment search tool. *J. Mol. Biol.* **1990**, *215*, 403–410. [CrossRef]
39. Felsenstein, J. Evolutionary trees from DNA sequences: A maximum likelihood approach. *J. Mol. Evol.* **1981**, *17*, 368–376. [CrossRef]
40. Kumar, S.; Stecher, G.; Li, M.; Knyaz, C.; Tamura, K. MEGA X: Molecular Evolutionary Genetics Analysis across Computing Platforms. *Mol. Biol. Evol.* **2018**, *35*, 1547–1549. [CrossRef]
41. Doussang, D.; Sallaberry-Pincheira, N.; Cabanne, G.S.; Lijtmaer, D.A.; González-Acuña, D.; Vianna, J.A. Specialist versus generalist parasites: The interactions between host diversity, environment and geographic barriers in avian malaria. *Int. J. Parasitol.* **2021**, *51*, 899–911. [CrossRef]
42. Beadell, J.S.; Covas, R.; Gebhard, C.; Ishtiaq, F.; Melo, M.; Schmidt, B.K.; Perkins, S.L.; Graves, G.R.; Fleischer, R.C. Host associations and evolutionary relationships of avian blood parasites from West Africa. *Int. J. Parasitol.* **2009**, *39*, 257–266. [CrossRef]
43. Illera, J.C.; López, G.; García-Padilla, L.; Moreno, Á. Factors governing the prevalence and richness of avian haemosporidian communities within and between temperate mountains. *PLoS ONE* **2017**, *12*, e0184587. [CrossRef]
44. Keesing, F.; Holt, R.D.; Ostfeld, R.S. Effects of species diversity on disease risk. *Ecol. Lett.* **2006**, *9*, 485–498. [CrossRef]
45. Ellis, V.A.; Huang, X.; Westerdahl, H.; Jönsson, J.; Hasselquist, D.; Neto, J.M.; Nilsson, J.-Å.; Nilsson, J.; Hegemann, A.; Hellgren, O.; et al. Explaining prevalence, diversity and host specificity in a community of avian haemosporidian parasites. *Oikos* **2020**, *129*, 1314–1329. [CrossRef]
46. Ferraguti, M.; la Puente, J.M.-d.; Jiménez, M.Á.; Llorente, F.; Roiz, D.; Ruiz, S.; Soriguer, R.; Figuerola, J. A field test of the dilution effect hypothesis in four avian multi-host pathogens. *PLoS Pathog.* **2021**, *17*, e1009637. [CrossRef]
47. Pinheiro, R.B.P.; Félix, G.M.F.; Chaves, A.V.; Lacorte, G.A.; Santos, F.R.; Braga, É.M.; Mello, M.A.R. Trade-offs and resource breadth processes as drivers of performance and specificity in a host-parasite system: A new integrative hypothesis. *Int. J. Parasitol.* **2016**, *46*, 115–121. [CrossRef]
48. Civitello, D.J.; Cohen, J.; Fatima, H.; Halstead, N.T.; Liriano, J.; McMahon, T.A.; Ortega, C.N.; Sauer, E.L.; Sehgal, T.; Young, S.; et al. Biodiversity inhibits parasites: Broad evidence for the dilution effect. *Proc. Natl. Acad. Sci. USA* **2015**, *112*, 8667–8671. [CrossRef]
49. Wood, C.L.; Lafferty, K.D.; DeLeo, G.; Young, H.S.; Hudson, P.J.; Kuris, A.M. Does biodiversity protect humans against infectious disease? *Ecology* **2014**, *95*, 817–832. [CrossRef]

50. Roiz, D.; Vázquez, A.; Ruiz, S.; Tenorio, A.; Soriguer, R.; Figuerola, J. Evidence that Passerine Birds Act as Amplifying Hosts for Usutu Virus Circulation. *Ecohealth* **2019**, *16*, 734–742. [CrossRef]
51. Chaisi, M.E.; Osinubi, S.T.; Dalton, D.L.; Suleman, E. Occurrence and diversity of avian haemosporidia in Afrotropical landbirds. *Int. J. Parasitol. Parasites Wildl.* **2019**, *8*, 36–44. [CrossRef]
52. Outlaw, D.C.; Harvey, J.A.; Drovetski, S.V.; Voelker, G. Diversity and distribution of avian haemosporidians in sub-Saharan Africa: An inter-regional biogeographic overview. *Parasitology* **2017**, *144*, 394–402. [CrossRef]
53. Loiseau, C.; Iezhova, T.; Valkiūnas, G.; Chasar, A.; Hutchinson, A.; Buermann, W.; Smith, T.B.; Sehgal, R.N.M. Spatial variation of haemosporidian parasite infection in African rainforest bird species. *J. Parasitol.* **2010**, *96*, 21–29. [CrossRef]
54. Wardjomto, M.B.; Ndlovu, M.; Pérez-Rodríguez, A.; Pori, T.; Nangammbi, T. Avian haemosporidia in native and invasive sparrows at an Afrotropical region. *Parasitol. Res.* **2021**, *120*, 2631–2640. [CrossRef]
55. Fecchio, A.; Bell, J.A.; Bosholn, M.; Vaughan, J.A.; Tkach, V.V.; Lutz, H.L.; Cueto, V.R.; Gorosito, C.A.; González-Acuña, D.; Stromlund, C.; et al. An inverse latitudinal gradient in infection probability and phylogenetic diversity for Leucocytozoon blood parasites in New World birds. *J. Anim. Ecol.* **2020**, *89*, 423–435. [CrossRef]
56. Musa, S.; Mackenstedt, U.; Woog, F.; Dinkel, A. Avian malaria on Madagascar: Prevalence, biodiversity and specialization of haemosporidian parasites. *Int. J. Parasitol.* **2019**, *49*, 199–210. [CrossRef]
57. Agosta, S.J.; Janz, N.; Brooks, D.R. How specialists can be generalists: Resolving the "parasite paradox" and implications for emerging infectious disease. *Zoologia* **2010**, *27*, 151–162. [CrossRef]
58. Johnson, K.P.; Malenke, J.R.; Clayton, D.H. Competition promotes the evolution of host generalists in obligate parasites. *Proc. R. Soc. B Biol. Sci.* **2009**, *276*, 3921–3926. [CrossRef]
59. Atkinson, C.T.; LaPointe, D.A. Introduced Avian Diseases, Climate Change, and the Future of Hawaiian Honeycreepers. *J. Avian Med. Surg.* **2009**, *23*, 53–63. [CrossRef]

animals

MDPI

Article

Differentiation of Synanthropic Fleas from Andalusia (Spain) through Geometric Morphometrics Analysis

Angela M. García-Sánchez, Ignacio Trujillo, Antonio Zurita * and Cristina Cutillas

Department of Microbiology and Parasitology, Faculty of Pharmacy, University of Seville, Profesor García González 2, 41012 Sevilla, Spain; agarcia77@us.es (A.M.G.-S.); nachotr95@hotmail.com (I.T.); cutillas@us.es (C.C.)
* Correspondence: azurita@us.es

Simple Summary: Fleas are blood-sucking insects that are not only a nuisance but can also act as vectors for various diseases in animals and humans, including dangerous ailments such as the bubonic plague. Identifying and classifying these insects accurately is crucial for understanding how they spread and how to control them. Geometric morphometrics, a cutting-edge technique, is proving to be an invaluable tool in this regard, alongside traditional methods and molecular biology. In the present study conducted in Andalusia, Spain, this technique successfully differentiated between three populations of fleas, providing insights into their distribution, size, and characteristics. Image processing software was employed to obtain measurements, such as perimeters and areas, of the fleas under study. These findings underscore the importance of geometric morphometrics in studying and managing arthropod populations, particularly in cases where other methods fall short or are not available.

Abstract: Fleas (Siphonaptera) are ectoparasitic hematophagous insects responsible for causing bites and itchy skin conditions in both humans and animals. Furthermore, they can act as vectors of different pathogens of a wide variety of diseases worldwide, including bartonellosis, rickettsiosis, and bubonic plague. Accurate identification of fleas is necessary for the study of their epidemiology, prevention, and control. In addition to traditional morphological classification approaches and molecular biology techniques, geometric morphometrics is increasingly proving to be a useful complementary tool for discriminating between Siphonaptera taxa. With the objective of determining the capacity of this technique to identify and differentiate synanthropic fleas, a principal component analysis was carried out on populations of *Ctenocephalides felis*, *Pulex irritans*, and *Archaeopsylla erinacei* collected in distinct regions of Andalusia (Spain). The analysis carried out on 81 male and female specimens revealed factorial maps that allowed the differentiation of the populations under study, with only partial overlaps that did not prevent their correct identification. Global size differences were also detected, with a slightly larger size in *P. irritans* males and a bigger size in *A. erinacei* females. Therefore, the present study emphasizes the role of geometric morphometrics as a useful complementary technique in taxonomic studies of arthropods, especially in the case of flea specimens lacking representative morphological features.

Keywords: Siphonaptera; *Ctenocephalides*; *Pulex*; *Archaeopsylla*; morphometrics

1. Introduction

Fleas (Siphonaptera) comprise a highly specialized order of holometabolous ectoparasitic insects with a cosmopolitan distribution and about 2700 species described so far [1]. In addition to being able to provoke bites and pruritic welts on the skin, these arthropods are also known to be vectors of different pathogens, responsible for causing a wide variety of diseases worldwide, including bartonellosis, rickettsiosis, and bubonic plague [2–9]. This is due to the nature of some flea species, which present a low host specificity that facilitates

the exchanging of microorganisms, posing a potential threat to the health of both humans and animals [10–12].

The prevention and control of fleas require a large investment of money per year, which represents a significant economic burden [13]. It is crucial to enhance our knowledge about the taxonomy of fleas to develop effective strategies to reduce flea infestations and their negative impact on our environment.

Recently, in addition to the traditional morphological identification and molecular biology approaches, geometric morphometrics has proven to be a useful complementary technique for discriminating taxa across different groups [14–16]. One of the main features of geometric morphometrics is that it is especially helpful in cases of taxa that present morphological ambiguity [15], a situation relatively common in fleas [17–19]. This scenario invited the exploration of the affordable criterion offered by geometric morphometrics in systematic studies on flea genera, with promising results in *Ctenocephalides* Stiles & Collins, 1930 [20,21], *Ctenophthalmus* Kolenati, 1856 [22], *Pulex* Linnaeus, 1758 [23], and *Stenoponia* Jordan & Rothschild, 1911 [24].

In Europe, there is evidence of an escalating frequency of vector-borne diseases and heightened pathogen circulation, primarily influenced by human-related factors [25]. One region remarkably affected by arthropod-borne diseases is Andalusia, situated in the southern of Spain, where the West Nile Virus circulation is more widespread than initially considered [26], and outbreaks have been reported in recent years [27].

Furthermore, the incidence of murine typhus, a zoonosis caused by *Rickettsia typhi* da Rocha Lima, 1916 transmitted to humans by fleas, seems to be increasing slowly in Andalusia [28], an aspect that reveals the ability of some flea species present in the region to transmit pathogenic bacteria. Hence, it is essential to resort to techniques that allow us to safely discern between taxa.

The cat flea, *Ctenocephalides felis* Bouché, 1835, succeeded in its expansion as a global parasite, and it is one of the most common flea species identified in domestic dogs and cats worldwide [11,19]. In Southwestern Europe, *C. felis* is the dominant species, although, in Eastern Europe, the infestation by *Ctenocephalides canis* Curtis, 1826 and *Pulex irritans* Linnaeus, 1758 also occurs [19,29–31].

On the other hand, hedgehogs inhabit rural, urban, and suburban environments, and they are frequently parasitized by blood-sucking arthropods, such as hard ticks and fleas, including the hedgehog flea, *Archaeopsylla erinacei* Bouché, 1835, and other flea species such as *C. felis*, *C. canis*, and *Nosopsyllus fasciatus* Bosc, 1800 [32–35]. Since they usually cohabit with pets and humans, they can potentially act as reservoirs of pathogen microorganisms responsible for zoonoses [12].

The main objective of the present study was to determine the capacity of geometric morphometric analysis to identify and discriminate fleas from populations of *C. felis*, *P. irritans*, and *A. erinacei* collected in Andalusia, in order to strengthen its role as a useful complementary technique in arthropod taxonomical studies.

2. Materials and Methods

2.1. Collection of Samples

Over a period of 19 months, we collected flea samples from dogs (*Canis lupus familiaris* Linnaeus, 1758) and one hedgehog (*Erinaceus europaeus* Linnaeus, 1758) that coexisted with other dogs.

To gather flea samples from the hosts, we reached out to some veterinary clinics, veterinary hospitals, pet shelters, and some pet owners. In total, we contacted 145 veterinary clinics and 30 pet shelters and kennels. Among these, 18 centers agreed to participate in the sample collection (see Acknowledgements). All participants volunteered for this sampling process. Only animals parasitized by fleas were sampled. Veterinary practitioners performed an initial inspection of pets brought to their facilities. Each pet was checked for fleas and examined by a veterinarian who recorded clinical signs related to flea infestation. Adult flea counts were conducted according to the World Association for the Advancement of Veterinary Parasitology guidelines [36]. In brief, the animals were combed over their entire bodies with a fine-toothed comb for 5–10 min.

All captured fleas from each infested host were transferred to a plastic 1.5 mL tube containing 96% ethanol for subsequent identification and morphometrics analyses.

2.2. Morphological Identification and Metric Data Processing

For morphological analysis, all specimens were initially examined under an optical microscope for specific classification. Following this, the specimens were cleared using 10% KOH, prepared, and mounted on glass slides following conventional procedures with the EUKITT mounting medium (O. Kindler GmbH & Co., Freiburg, Germany) [37]. The cleared and mounted specimens were examined again for a more detailed morphological analysis using a BX61 microscope (Olympus, Tokyo, Japan) and submitted to image capture processes with the imaging software cellSens Standard version 4.2 (Olympus, Tokyo, Japan). The diagnostic morphological characters of all samples were analyzed by comparison with figures, keys, and descriptions reported previously [38–42]. The measurement images of each flea were made using the image analysis software Image-PRO v11 (Media Cybernetics, Rockville, MD, USA). A total of 28 different parameters were measured for males and 36 for females (Tables 1 and 2).

Table 1. Biometrical data of males of *C. felis*, *P. irritans*, and *A. erinacei* collected from *Canis lupus familiaris* and *Erinaceus europaeus* from Andalucia (Spain).

	C. felis					*P. irritans*					*A. erinacei*				
	MAX	MIN	Mean	SD	VC	MAX	MIN	Mean	SD	VC	MAX	MIN	Mean	SD	VC
Global measures															
Area (mm^2) †	1058.4	659.3	847.3	123.4	15	1379.8	688.6	1009.6	276	27	1436.7	990.3	1225.5	153.3	13
Roundness †	2.41	2.02	2.20	0.11	5	2.06	1.88	1.95	0.08	4	2.48	2.09	2.32	0.14	6
Circularity †	0.36	0.30	0.33	0.02	5	0.5	0.42	0.46	0.03	6	0.43	0.35	0.38	0.02	6
Perimeter (µm) †	5642.4	4213.9	4828.5	418.3	9	5676.4	4114.8	4925.4	649.5	13	6685.8	5424.4	5958	410.7	7
TL (µm)	2084	1563.0	1777.7	138.3	8	2027.7	1312.2	1650.0	285.8	17	2094.5	1667.3	1877.9	158.5	8
TW (µm) †	815.1	627.5	722.1	52.6	7	1064.1	772.3	926.0	120.9	13	989.4	810.1	890.0	56.4	6
Head measures															
Area (µm^2) †	78,287	53,982	68,194	6318	9	100,673	74,318	83,921	9520	11	145,712	114,111	127,334	11,327	9
Roundness †	1.39	1.29	1.35	0.02	2	1.31	1.23	1.27	0.03	2	1.24	1.17	1.19	0.02	2
Circularity †	0.46	0.42	0.44	0.01	3	0.50	0.45	0.48	0.02	4	0.63	0.59	0.61	0.02	3
Perimeter (µm) †	1153	953.8	1072.6	49.8	5	1261.0	1071.6	1153.9	68.6	6	1502.9	1302.7	1379.8	64.0	5
HL (µm) †	391.1	306.8	360	22.9	6	357.0	293.4	317.6	25.0	8	453.2	378.3	412.6	23.2	6
HW (µm) †	251.2	210.0	228.4	12.1	5	263.0	228.3	248.2	15.1	6	378.1	323.4	352.2	16.2	5
Prothorax measures															
Area (µm^2) †	24,848	14,287	19,192	3395	18	20,140	9635	13,081	3712	28	33,268	18,711	28,802	4038	14
Roundness †	1.45	1.23	1.37	0.06	4	2.08	1.68	1.90	0.14	7	1.60	1.25	1.43	0.11	8
Circularity †	0.58	0.46	0.51	0.04	8	0.36	0.26	0.32	0.04	11	0.61	0.38	0.46	0.07	16
Perimeter (µm) †	665.5	478.9	570.2	50.3	9	651.1	466.6	550.1	60.4	11	778.0	594.9	715.2	53.0	7
PROTW (µm) †	120.4	82.0	100.8	11.0	11	83.5	46.1	60.1	12.8	21	151.0	88.6	120.7	16.6	14
Mesothorax measures															
Area (µm^2) †	31,142	14,043	22,358	4518	20	28,396	8581	18,186	7347	40	33,772	23,289	28,643	3716	13
Roundness †	1.64	1.33	1.46	0.08	6	2.05	1.65	1.81	0.19	10	1.92	1.40	1.72	0.19	11
Circularity †	0.57	0.42	0.48	0.04	8	0.43	0.29	0.37	0.05	15	0.55	0.37	0.42	0.05	13
Perimeter (µm) †	723.9	512.8	634.1	63.3	10	764.6	467.7	624.0	107.4	17	858.8	670.8	783.0	64.7	8
MESOW (µm) †	137.4	88.7	109.0	13.8	13	103.5	47.7	76.2	21.8	29	137.0	99.3	111.3	13.6	12
Metathorax measures															
Area (µm^2) †	37,152	20,305	28,528	5061	18	53,522	20,280	35,569	12,210	34	48,486	33,659	38,673	4209	11
Roundness †	1.57	1.28	1.46	0.07	5	1.64	1.38	1.53	0.10	7	1.81	1.47	1.60	0.10	7
Circularity †	0.57	0.39	0.44	0.04	10	0.47	0.42	0.44	0.02	5	0.42	0.34	0.39	0.02	6
Perimeter (µm) †	877.8	625.5	758.6	71.6	9	1024.4	667.7	857.8	123.5	14	1023.2	883.7	922.7	41.9	5
METW (µm) †	142.8	102.3	120.4	12.6	10	177.1	97.5	131.4	30.4	23	157.1	109.7	134.2	13.9	10
AW (µm) †	34.9	12.9	20.7	6.0	29	51.3	19.2	37.0	12.2	33	31.3	11.3	23.1	6.2	27

TL = total length, TW = total width, HL = total length of the head, HW = total width of the head, PROTW = total width of the prothorax, MESOW = total width of the mesothorax, METW = total width of the metathorax, AW: apex width, MAX = maximum, MIN = minimum, SD = standard deviation, Mean = arithmetic mean, VC = coefficient of variation (percentage converted), † = significant differences between groups ($p < 0.005$).

Table 2. Biometrical data of females of *C. felis*, *P. irritans*, and *A. erinacei* collected from *Canis lupus familiaris* and *Erinaceus europaeus* from Andalucia (Spain).

	C. felis					*P. irritans*					*A. erinacei*				
	MAX	MIN	Mean	SD	VC	MAX	MIN	Mean	SD	VC	MAX	MIN	Mean	SD	VC
Global measures															
Area (mm^2)	2290.4	827.8	1724.7	438.2	25	2876.6	1180.3	1596.4	567.1	36	2874.6	1368.8	2023.6	434.1	21
Roundness †	2.57	1.82	2.10	0.20	10	2.11	1.80	1.90	0.08	4	2.34	1.76	2.02	0.17	8
Circularity †	0.40	0.33	0.36	0.02	5	0.50	0.40	0.46	0.02	5	0.49	0.36	0.41	0.03	8
Perimeter (µm) †	7967.9	4583.8	6661.7	837.0	13	8379.8	5310.4	6092.9	993.7	16	8733.5	6038.3	7127.7	902.6	13
TL (mm) †	2859.5	1616.2	2439.8	337.8	14	2861.0	1757.8	2041.3	372.8	18	2938.3	1885.7	2425.1	332.8	14
TW (mm) †	1208.6	767.4	1038.7	136.2	13	1495.9	1003.5	1150.2	144.0	13	1487.5	1058.3	1220.5	112.7	9
Head measures															
Area (µm^2) †	100,169	67,749	83,173	9264	11	135,023	87,601	111,137	12,669	11	163,067	107,663	135,800	18,243	13
Roundness †	1.55	1.35	1.45	0.05	3	1.40	1.22	1.30	0.06	4	1.26	1.16	1.20	0.02	2
Circularity †	0.46	0.38	0.42	0.02	5	0.50	0.43	0.46	0.02	5	0.58	0.51	0.55	0.02	4
Perimeter (µm) †	1354.0	1102.8	1229.3	73.2	6	1455.0	1238.6	1342.4	65.3	5	1564.9	1273.1	1423.1	89.7	6
HL (µm) †	449.5	357.9	406.6	27.3	7	428.5	344.0	371.5	24.4	7	463.8	380.5	421.3	26.4	6
HW (µm) †	296.8	243.2	265.4	15.1	6	330.1	243.7	287.6	22.2	8	383.9	274.2	324.7	30.5	9
Prothorax measures															
Area (µm^2) †	36,972	15,844	25,774	5347	21	29,118	16,764	22,842	3602	16	48,287	25,729	37,136	7542	20
Roundness †	1.92	1.40	1.65	0.14	8	2.22	1.53	1.78	0.18	10	1.50	1.26	1.42	0.08	5
Circularity †	0.53	0.36	0.45	0.05	10	0.43	0.26	0.34	0.04	13	0.59	0.41	0.48	0.05	11
Perimeter (µm) †	876.7	576.8	723.8	74.3	10	812.5	615.0	709.3	51.4	7	879.0	691.0	805.6	71.8	9
PROTW (µm) †	139.0	67.7	105.0	18.3	17	98.6	56.9	84.8	12.2	14	168.1	106.3	138.2	20.2	15
Mesothorax measures															
Area (µm^2)	48,419	20,032	35,002	8398	24	49,243	27,112	37,914	5734	15	50,800	29,397	40,732	6543	16
Roundness †	1.69	1.24	1.51	0.11	7	1.86	1.52	1.64	0.10	6	1.86	1.50	1.60	0.12	7
Circularity †	0.60	0.33	0.46	0.06	13	0.46	0.35	0.42	0.03	7	0.51	0.40	0.47	0.03	6
Perimeter (µm) †	967.0	638.1	807.6	101.4	13	1031.0	732.8	879.5	78.2	9	1012.9	783.6	899.3	75.2	8
MESOW (µm) †	161.4	88.1	135.2	19.0	14	145.5	105.9	124.5	10.0	8	168.1	113.8	140.8	15.6	11
Metathorax measures															
Area (µm^2) †	63,421	26,207	46,240	10,448	23	73,621	48,603	63,117	7854	12	66,054	44,555	54,053	7425	14
Roundness †	2.09	1.41	1.68	0.17	10	1.73	1.37	1.51	0.10	7	1.59	1.31	1.48	0.07	5
Circularity †	0.48	0.34	0.41	0.03	9	0.48	0.37	0.44	0.03	8	0.50	0.40	0.44	0.03	7
Perimeter (µm) †	1185.4	772.5	979.9	128.9	13	1199.2	943.2	1090.1	70.3	6	1116.9	897.5	997.9	73.2	7
METW (µm) †	175.8	109.0	147.5	18.5	13	206.9	145.3	178.4	15.8	9	188.3	144.9	170.2	16.1	9
Spermatheca measures															
Area (µm^2) †	3088	2162	2619	259	10	3489	2439	3062	309	10	11,768	3875	7474	2120	28
Roundness †	1.21	1.05	1.14	0.04	4	1.06	1.01	1.03	0.01	1	1.25	1.05	1.12	0.06	6
Circularity †	0.76	0.59	0.65	0.05	7	0.89	0.76	0.85	0.04	5	0.81	0.53	0.66	0.08	12
BULGAP (µm) †	212.1	167.8	192.7	10.9	6	210.8	176.3	198.0	10.3	5	396.7	227.0	320.4	45.6	14
BULGAL (µm) †	71.5	52.6	61.2	5.0	8	70.6	54.3	63.9	4.7	7	127.0	80.3	99.7	15.0	15
BULGAW (µm) †	58.2	46.4	50.9	3.0	6	66.7	55.6	59.6	3.1	5	133.7	57.8	90.8	22.6	25
APEHILL (µm) †	52.8	22.9	35.6	9.1	25	64.2	26.9	46.9	12.2	26	118.9	38.4	83.4	21.0	25
APEHILW (µm) †	31.7	17.0	24.7	4.4	18	64.5	23.4	31.7	10.7	34	57.9	20.9	38.1	9.4	25
DBMV (µm) †	349	131	251	57.1	23	485	290	378	56.7	15	503.8	150.3	293.4	116.5	40

TL = total length, TW = total width, HL = total length of the head, HW = total width of the head, PROTW = total width of the prothorax, MESOW = total width of the mesothorax, METW = total width of the metathorax, BULGAP: perimeter of the bulga, BULGAL: total length of the bulga, BULGAW: total width of the bulga, APEHILL: total length of the apex of the hilla, APEHILW: total width of the apex of the hilla, DBMV = distance from bulga to the ventral margin of the body, MAX = maximum, MIN = minimum, SD = standard deviation, Mean = arithmetic mean, VC = coefficient of variation (percentage converted), † = significant differences between groups ($p < 0.005$).

Descriptive univariate statistics based on arithmetic mean, standard deviation, and coefficient of variation for all parameters were determined for male and female populations. The data were subjected to one-way ANOVA (analysis of variance) for statistical analysis of the parameters. The results were statistically significant when $p < 0.05$. Statistical analysis was performed using Microsoft Excel for Microsoft 365 MSO (v2402). In addition, biometric characters of fleas were compared between different species and the most significant parameters were assayed for a morphometrics study.

Morphological variation was quantified using geometric morphometrics [43], a technique that provides an estimate of size integrating different growth axes into a single variable known as "centroid size" [44]. The estimate of size was represented by a single variable that reflected variation in multiple directions, as many as there were landmarks under study, and shape was defined as their relative positions after correction for size, position, and orientation. With these informative data, and the corresponding software freely available to conduct complex analyses, significant biological and epidemiological features can be quantified more accurately [45].

Multivariate analyses were applied to assess phenotypic variations among the samples, using size-free canonical discriminant analysis on the covariance of log-transformed measurements. These analyses are applied to exclude the effect of within-group ontogenetic variations by reducing the effect of each character on the first pooled within-group principal component (a multivariate size estimator) [46]. principal component analysis (PCA) was used to summarize most of the variations in a multivariate dataset in a few dimensions [47]. Morphometric data were explored using multivariate analysis in three parameters in males (TW, HW, and AW) (Table 1) and females (Global Circularity, BULGAP, and HL) in females (Table 2) using BAC v.2 software [21,48].

3. Results

A total of 81 fleas (34 males and 47 females) were collected from different regions of Andalusia and classified as follows: 39 as *C. felis* (18 males and 21 females), 19 as *P. irritans* (6 males and 13 females), and 23 as *A. erinacei* (10 males and 13 females) (Table 3). All *C. felis* specimens and three *P. irritans* (the only male from Huelva and two female fleas from Seville) were collected from dogs (*Canis lupus familiaris*), whereas *A. erinacei* were collected from a hedgehog host. The rest of the *P. irritans* fleas were collected off-host from a neglected horse stable (Table 3).

Table 3. Distribution of fleas collected from dogs from different geographical origins.

Geographical Origin	*C. felis* (Number of Fleas)		*P. irritans* (Number of Fleas)		*A. erinaceid* (Number of Fleas)	
	Male	**Female**	**Male**	**Female**	**Male**	**Female**
Sanlúcar de Barrameda (Cadiz, Spain)	18	21	-	-	-	-
Seville (Seville, Spain)	-	-	5	13	-	-
Huelva (Huelva, Spain)	-	-	1	-	-	-
Dos Hermanas (Seville, Spain)	-	-	-	-	10	13
Total	18	21	6	13	10	13

To carry out the classification of the samples, we considered descriptions used traditionally to discern between these species and, additionally, remarkable morphological features based on the measurements performed. Statistical tests showed several significant measurements for subsequent morphometric analyses. Therefore, the following parameters were used: total width (TW), total width of the head (HW), and apex width (AW) in males (Table 1) and total length of the head (HL), perimeter of the bulga (BULGAP), and Global Circularity in females (Table 2). This perimeter is the length of the bulga's boundary, whereas the circularity is calculated as the ratio of the area of an object against a circle whose diameter is equal to the object's maximum feret. The influence of size was analyzed using PCA in *C. felis*, *P. irritans*, and *A. erinacei*, involving the regression of each character separately on the within-group first principal component (PC1). The resulting factor maps for male and female populations are shown in Figures 1 and 2, respectively.

Male variables significantly correlated with PC1, contributing 71% to the overall variation. The male factor maps showed global size differences in the flea populations, with a slightly larger size in *P. irritans* males (Figure 1). The three male communities are well grouped in the factor map, with a lack of noteworthy overlapping areas between them. Only *C. felis* and *P. irritans* showed a partial overlap but with no inconvenience in their identification.

On the other hand, female variables significantly correlated with PC1, contributing 90% to the overall variation. The resulting factor maps (Figure 2) clearly illustrate global size differences in the populations analyzed, including a bigger size in *A. erinacei*. As in the previous factor maps, there is a lack of notable overlapping areas between the female populations. As in the male factor maps, *C. felis* and *P. irritans* showed a partial overlap that did not prevent their identification.

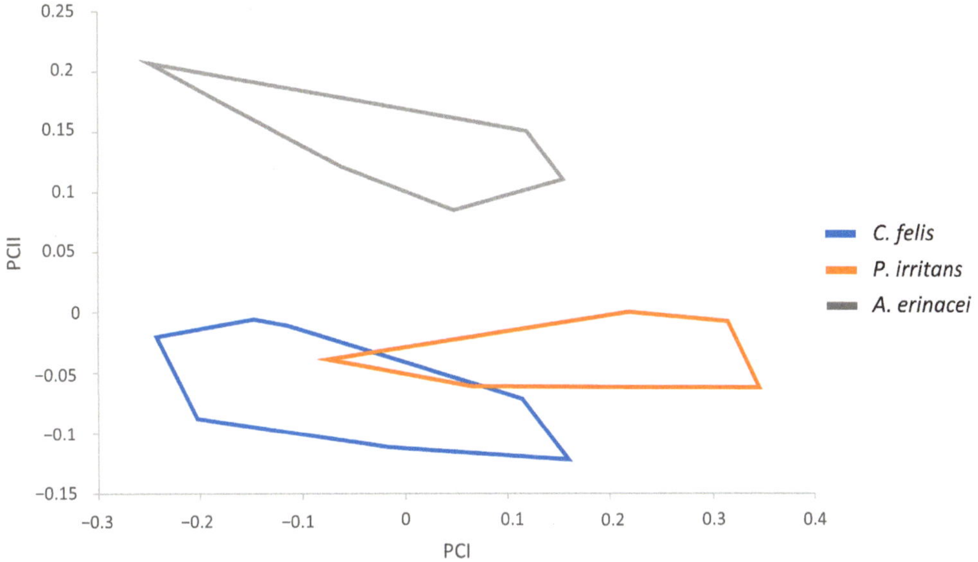

Figure 1. Factor map corresponding to *C. felis*, *P. irritans*, and *A. erinacei* male adults. Samples are projected onto the first and second principal components: PC1 (71%) and PC2 (26%). Each group is represented by its perimeter.

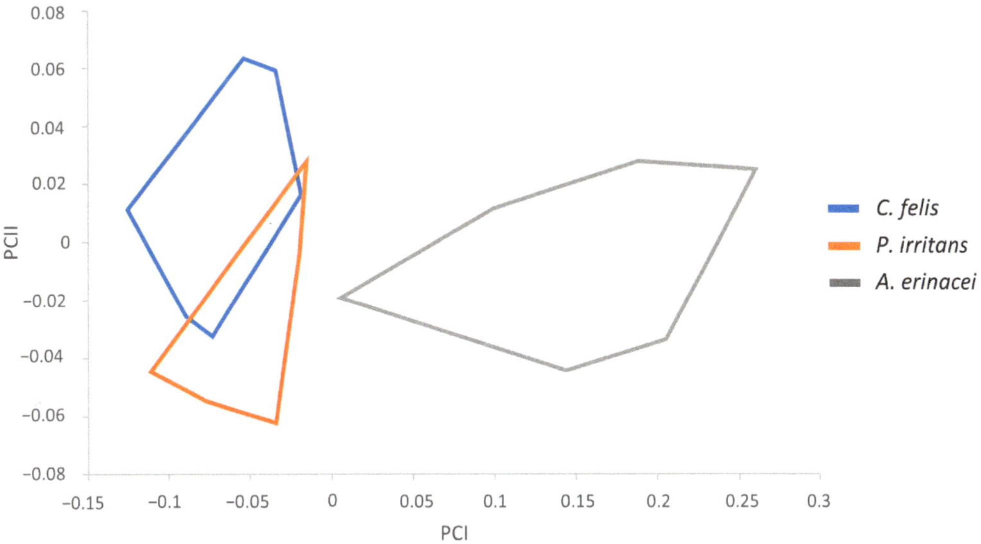

Figure 2. Factor map corresponding to *C. felis*, *P. irritans*, and *A. erinacei* female adults. Samples are projected onto the first and second principal components: PC1 (90%) and PC2 (8%). Each group is represented by its perimeter.

4. Discussion

The accurate classification of fleas requires careful morphological examination or molecular confirmation, and therefore, the possibility that prior studies may have inadvertently misidentified fleas cannot be discarded [19,49]. In fact, authors like Ménier and Beaucournu reported numerous misidentifications in specimens of the genus *Ctenocephalides* [50]. The study of flea epidemiology, control, and prevention requires the accurate identification of species and subspecies.

In general, the classification of genera and species of fleas is based on external morphological characters. The presence or absence of combs and eyes, along with the length of the head, are typically significant features in morphological identification [51]. However, the size itself could never represent a way to reliably recognize the sex or the species of a flea specimen [52], and some flea species do not have easily identifiable morphological characters. For instance, *A. erinacei* and *P. irritans* do not possess pronotal or genal combs, so their classification process can be more complex in case the required specialized skills in flea identification are lacking. It is necessary to pay great attention to detail to recognize developmental stages and adequate sex identification [52]. Additionally, *C. felis* is a good example of a flea species known for its morphological ambiguity and the underlying issues in the study of their global populations [19]. In terms of molecular biology techniques, the notable lack of large-scale phylogenetic data for flea taxa causes some genera like *Ctenocephalides* to not have a defined genetic identity [17,19,40], especially if we consider subspecies [19].

Furthermore, it is still surprising that, despite the considerable veterinary and public health significance of dog fleas, studies investigating the diversity of these ectoparasite species on pets and the occurrence of flea-borne pathogens are scarce in certain regions [20].

Given the difficulties associated with flea morphological identification, the limited genetic information available, and the insufficient knowledge of the common pathogens of each flea species, the need to resort to complementary diagnosis techniques arises.

Geometric morphometrics analysis is one of these novel approaches applied to parasitological diagnosis, usually employed in arthropod identification [43]. The technique is based on the utilization of computer software for data processing and interpretation, with the advantage being that costly reagents and equipment are not required. Its affordability and the simplicity of data collection make it especially useful in low-resource settings [52,53].

To the best of our knowledge, the present work is the first in which flea measurements have been obtained using the imaging software Image-PRO. This program has been used before to study arthropods and other parasites [54–57]. It allowed us to incorporate into the analysis accurate measures such as areas, perimeters and circularities of the flea specimens for the first time. Global Circularity and BULGAP were revealed as useful features that contribute to the identification of flea species via geometric morphometrics.

The distribution of the three analyzed flea species showed a comparable pattern in both factor maps. *A. erinacei* appeared distant from *C. felis* and *P. irritans*, showing an appreciably larger size in the case of females. *C. felis* and *P. irritans* presented a small overlapping area, which did not prevent their individual identification. In both cases, *P. irritans* has always appeared larger than *C. felis*.

Although *A. erinacei* and *C. felis* are part of the Pulicinae family, the factor maps illustrated that fleas at the same taxonomic level are not necessarily closer at the morphological level since *A. erinacei* appears further away from both *C. felis* and *P. irritans*.

The selection of representative measurements for the morphometrics analyses was in accordance with previously published works. Total width (TW), head width (HW), and apex width (AW) are consolidated as useful parameters that define the morphological identity in males [21,50,58], as well as the total length of the head (HL) and the perimeter of the bulga (BULGAP) for females [21]. Due to the lack of genal ctenidium in *P. irritans* and *A. erinacei*, the difference in length between first and second spines (DEG parameter [21]) could not be considered in the present analysis, whereas the inclusion of the degree of elongation of the apical part (hilla) in females offered similar results but with a bit more overlap between *P. irritans* and *C. felis*. This is why the parameter APEHILL [21] was substituted by BULGAP, which permits the best differentiation between species.

The three analyzed flea species are among the most frequent in our environment, and, as a result, there is a notable risk of encountering them, with consequently associated parasitism suffered by humans and animals. After applying geometric morphometrics to differentiate flea communities of the same genera [21–24], the present work represents a step further, since this technique allowed us to identify different flea genera.

5. Conclusions

Accurate identification of fleas is necessary for studying the epidemiology, prevention, and control of these arthropods. In situations of uncertainty, alternative approaches are required to ensure correct classification. Geometric morphometrics is increasingly recognized as a reliable complementary technique for identifying flea species, particularly valuable in environments with limited resources.

In the present work, we were able to discern between the flea species *A. erinacei*, *P. irritans*, and *C. felis* using principal component analysis of males and females. Differences in overall size were also detected: *A. erinacei* presented the largest size in females, whereas *P. irritans* was slightly larger in males. Therefore, morphometrics is a relevant technique with great potential for application in the field of fleas, considering the existence of species that have traditionally posed challenges in their identification.

Author Contributions: Conceptualization, C.C. and A.Z.; methodology, A.M.G.-S.; software, A.M.G.-S. and I.T.; validation, A.Z. and C.C.; formal analysis, C.C.; investigation, A.M.G.-S. and I.T.; resources, C.C.; data curation, A.M.G.-S.; writing—original draft preparation, A.M.G.-S.; writing—review and editing, A.Z. and C.C.; visualization, C.C.; supervision, A.Z. and C.C.; project administration, C.C.; funding acquisition, C.C. All authors have read and agreed to the published version of the manuscript.

Funding: This research was funded by a project from the Junta de Andalucía, European Regional Development Fund (FEDER) (Ref. P20_00544).

Institutional Review Board Statement: Not applicable.

Informed Consent Statement: Not applicable.

Data Availability Statement: The data presented in this study are available in the article.

Acknowledgments: We would like to express our gratitude to these participating veterinary clinics, pet shelters, and kennels for their continued support and timely submission of samples, which greatly contributed to the success of our study. Veterinary clinics: *El acuario*, Ayamonte, Huelva; *Animal-Vetx El Saladillo*, Huelva; *Cuenca minera*, Minas de Río Tinto, Huelva; *Consultorio veterinario Galaroza*, Galaroza, Huelva; *Roncal*, Pilas, Sevilla; *Mascotasalud*, Tomares, Sevilla; *Ronda de capuchinos*, Sevilla; *Vilovet*, Cazalla de la Sierra, Sevilla; *Taxana*, Los Rosales, Sevilla; *Sobrino E Illescas*, Ecija, Sevilla; *SyR*, Estepa, Sevilla; *Alfavet*, Fuente Palmeras, Córdoba; *Mascovet*, Posadas, Córdoba; *El Pinar*, Puerto de Santa María, Cádiz; *El Parque*, San Fernando, Cádiz. Pet shelters and kennels: *Mascotomares*, Tomares, Sevilla; *Arca*, Sevilla; *Sociedad protectora de animales y plantas*, Sevilla.

Conflicts of Interest: The authors declare no conflicts of interest.

References

1. Hastriter, M.W.; Bossard, R.L.; Lewis, R.E. Flea (Siphonaptera) World Species List (Spreadsheet). 2018. Available online: https://esanetworks.org/group/fleanews/page/flea-species-of-theworld-spreadsheetupdated-3-february-2018 (accessed on 3 December 2023).
2. Fleta Zaragozano, J. Rickettsiosis transmitidas por piojos, pulgas y ácaros. *Med. Integr.* **2002**, *39*, 147–152.
3. Parola, P.; Davoust, B.; Raoult, D. Tick- and flea-borne rickettsial emerging zoonoses. *Vet. Res.* **2005**, *36*, 469–492. [CrossRef] [PubMed]
4. De Sousa, R.; Fournier, P.E.; Santos-Silva, M.; Amaro, F.; Bacellar, F.; Raoult, D. Molecular detection of *Rickettsia felis, Rickettsia typhi* and two genotypes closely related to *Bartonella elizabethae*. *Am. J. Trop. Med. Hyg.* **2006**, *75*, 727–731. [CrossRef] [PubMed]

5. Beliavskaia, A.; Tan, K.K.; Sinha, A.; Husin, N.A.; Lim, F.S.; Loong, S.K.; Bell-Sakyi, L.; Carlow, C.K.S.; Abubakar, S.; Darby, A.C.; et al. Metagenomics of culture isolates and insect tissue illuminate the evolution of *Wolbachia*, *Rickettsia* and *Bartonella* symbionts in *Ctenocephalides* spp. Fleas. *Microb. Genom.* **2023**, *9*, 001045. [CrossRef] [PubMed]
6. Boyer, S.; Gillespie, T.R.; Miarinjara, A. *Xenopsylla cheopis* (rat flea). *Trends Parasitol.* **2022**, *38*, 607–608. [CrossRef] [PubMed]
7. Millán, J.; Sepúlveda-García, P.; Di Cataldo, S.; Canales, N.; Sallaberry-Pincheira, N.; Painean, J.; Cevidanes, A.; Müller, A. Molecular identification of *Bartonella* spp. and *Rickettsia felis* in fox fleas; Chile. *Comp. Immunol. Microb.* **2023**, *96*, 101983. [CrossRef] [PubMed]
8. Sepúlveda-García, P.; Alabi, A.; Álvarez, K.; Rojas, L.; Mella, A.; Gonçalves, L.R.; André, M.R.; Machado, R.; Müller, A.; Monti, G. *Bartonella* spp. in households with cats: Risk factors for infection in cats and human exposure. *One Health* **2023**, *16*, 100545. [CrossRef]
9. Yomogida, K.; Kjemtrup, A.; Martínez-López, B.; Ibrahim, M.; Contreras, Z.; Ngo, V.; Halai, U.A.; Balter, S.; Feaster, M.; Zahn, M.; et al. Surveillance of Flea-Borne Typhus in California; 2011–2019. *Am. J. Trop. Med. Hyg.* **2024**, *110*, 142–149. [CrossRef] [PubMed]
10. Kernif, T.; Socolovschi, C.; Wells, K.; Lakim, M.B.; Inthalad, S.; Slesak, G.; Boudebouch, N.; Beaucournu, J.C.; Newton, P.N.; Raoult, D.; et al. *Bartonella* and *Rickettsia* in arthropods from the Lao PDR and from Borneo; Malaysia. *Comp. Immunol. Microbiol. Inf. Dis.* **2012**, *35*, 51–57. [CrossRef]
11. Clark, N.J.; Seddon, J.M.; Šlapeta, J.; Wells, K. Parasite spread at the domestic animal-wildlife interface: Anthropogenic habitat use, phylogeny and body mass drive risk of cat and dog flea (*Ctenocephalides* spp.) infestation in wild mammals. *Parasit. Vectors.* **2018**, *11*, 8. [CrossRef]
12. Benkacimi, L.; Diarra, A.Z.; Bompar, J.M.; Bérenger, J.M.; Parola, P. Microorganisms associated with hedgehog arthropods. *Parasit. Vectors* **2023**, *16*, 211. [CrossRef]
13. Blagburn, B.L.; Dryden, M.W. Biology; treatment; and control of flea and tick infestations. *Vet. Clin. N. Am. Small Anim.* **2009**, *39*, 1173–1200. [CrossRef] [PubMed]
14. Villemant, C.; Simbolotti, G.; Kenis, M. Discrimination of *Eubazus* (Hymenoptera; Braconidae) sibling species using geometric morphometrics analysis of wing venation. *Syst. Biol.* **2007**, *32*, 625–634. [CrossRef]
15. Ruane, S. Using geometric morphometrics for integrative taxonomy: An examination of head shapes of milksnakes (genus *Lampropeltis*). *Zool. J. Linn. Soc.* **2015**, *174*, 394–413. [CrossRef]
16. Li, S.; Ricchiardi, E.; Bai, M.; Yang, X. A taxonomy review of *Oreoderus* Burmeister, 1842 from China with a geometric morphometric evaluation (Coleoptera; Scarabaeidae; Valgini). *ZooKeys* **2016**, *552*, 67–89.
17. Beaucournu, J.C.; Ménier, K. Le genre *Ctenocephalides* Stiles et Collins; 1930 (Siphonaptera; Pulicidae). *Parasite* **1998**, *5*, 3–16. [CrossRef]
18. Zurita, A.; Callejón, R.; de Rojas, M.; Cutillas, C. Morphological and molecular study of the genus *Nosopsyllus* (Siphonaptera: Ceratophyllidae). *Nosopsyllus barbarus* (Jordan & Rothschild 1912) as a junior synonym of *Nosopsyllus fasciatus* (Bosc d'Antic; 1800). *Insect Syst. Evol.* **2018**, *49*, 81–101.
19. Lawrence, A.L.; Webb, C.E.; Clark, N.J.; Halajian, A.; Mihalca, A.D.; Miret, J.; D'Amico, G.; Brown, G.; Kumsa, B.; Modrý, D.; et al. Out-of-Africa, human-mediated dispersal of the common cat flea, *Ctenocephalides felis*: The hitchhiker's guide to world domination. *Int. J. Parasitol.* **2019**, *49*, 321–336. [CrossRef]
20. Deak, G.; Safarov, A.; Xie, X.C.; Wang, R.; Mihalca, A.D.; Šlapeta, J. Fleas from the Silk Road in Central Asia: Identifcation of *Ctenocephalides canis* and *Ctenocephalides orientis* on owned dogs in Uzbekistan using molecular identification and geometric morphometrics. *Parasite. Vectors* **2022**, *15*, 345. [CrossRef]
21. García-Sánchez, A.M.; Zurita, A.; Cutillas, C. Morphometrics as a Complementary Tool in the Differentiation of Two Cosmopolitan Flea Species: *Ctenocephalides felis* and *Ctenocephalides canis*. *Insects* **2022**, *13*, 707. [CrossRef]
22. Zurita, A.; García-Sánchez, Á.M.; Cutillas, C. *Ctenophthalmus baeticus boisseauorum* (Beaucournu; 1968) and *Ctenophthalmus apertus allani* (Smit; 1955) (Siphonaptera: Ctenophthalmidae) as synonymous taxa: Morphometric; phylogenetic; and molecular characterization. *Bull. Entomol. Res.* **2020**, *110*, 663–676. [CrossRef]
23. Zurita, A.; Callejón, R.; García-Sánchez, Á.M.; Urdapilleta, M.; Lareschi, M.; Cutillas, C. Origin, evolution, phylogeny and taxonomy of *Pulex irritans*. *Med. Vet. Entomol.* **2019**, *33*, 296–311. [CrossRef] [PubMed]
24. Zurita, A.; García-Sánchez, Á.M.; Cutillas, C. Comparative molecular and morphological study of *Stenoponia tripectinata tripectinata* (Siphonaptera: Stenoponiidae) from the Canary Islands and Corsica. *Bull. Entomol. Res.* **2022**, *112*, 681–690. [CrossRef] [PubMed]
25. Beugnet, F.; Marié, J.L. Emerging arthropod-borne diseases of companion animals in Europe. *Vet. Parasitol.* **2009**, *163*, 298–305. [CrossRef]
26. García-Bocanegra, I.; Arenas-Montes, A.; Napp, S.; Jaén-Téllez, J.A.; Fernández-Morente, M.; Fernández-Molera, V.; Arenas, A. Seroprevalence and risk factors associated to West Nile virus in horses from Andalusia; Southern Spain. *Vet. Microbiol.* **2012**, *160*, 341–346. [CrossRef]
27. Hiraldo, J.D.G.; Fuerte-Hortigón, A.; Domínguez-Mayoral, A.; De la Rosa Riestra, S.; Palacios-Baena, Z.R.; Fernández, F.S.; Ruiz, R.L.; Pascual-Vaca, D.; de León, C.M.; Hurtado, R.J.; et al. Uncovering the neurological effects of West Nile virus during a record-breaking southern Spain outbreak in 2020–2021. *J. Neuroimmunol.* **2023**, *383*, 578179. [CrossRef]
28. Rodríguez-Alonso, B.; Almeida, H.; Alonso-Sardón, M.; Velasco-Tirado, V.; Robaina Bordón, J.M.; Carranza Rodríguez, C.; Pérez Arellano, J.L.; Belhassen-García, M. Murine typhus. How does it affect us in the 21st century? The epidemiology of inpatients in Spain (1997–2015). *Int. J. Infect. Dis.* **2020**, *96*, 165–171. [CrossRef]

29. Beugnet, F.; Labuschagne, M.; Fourie, J.; Jacques, G.; Farkas, R.; Cozma, V.; Halos, L.; Hellmann, K.; Knaus, M.; Rehbein, S. Occurrence of *Dipylidium caninum* in fleas from client-owned cats and dogs in Europe using a new PCR detection assay. *Vet. Parasitol.* **2014**, *205*, 300–306. [CrossRef] [PubMed]
30. Zurita, A.; Benkacimi, L.; El Karkouri, K.; Cutillas, C.; Parola, P.; Laroche, M. New records of bacteria in different species of fleas from France and Spain. *Comp. Immunol. Microbiol. Inf. Dis.* **2021**, *76*, 101648. [CrossRef]
31. Zurita, A.; Trujillo, I.; García-Sánchez, Á.M.; Cutillas, C. Survey of flea infestation in cats and dogs in Western Andalusia; Spain: Seasonality and other risk factors for flea infestation. *Med. Vet. Entomol.* **2024**, *38*, 244–251. [CrossRef]
32. Beck, W.; Clark, H.H. Differentialdiagnose medizinisch relevanter Flohspezies und ihre Bedeutung in der Dermatologie. *Der Hautarzt* **1997**, *48*, 714–719. [CrossRef] [PubMed]
33. Beck, W. Animal parasites and dermatophytes as a cause of epizoonoses in humans. *Der Prakt. Tierarzt* **2005**, *86*, 426–434.
34. Gaglio, G.; Allen, S.; Bowden, L.; Bryant, M.; Morgan, E.R. Parasites of European hedgehogs (*Erinaceus europaeus*) in Britain: Epidemiological study and coprological test evaluation. *Eur. J. Wildl. Res.* **2010**, *56*, 839–844. [CrossRef]
35. Dudek, K.; Földvári, G.; Majláthová, V.; Majláth, I.; Rigó, K.; Molnár, V.; Tóth, M.; Jankowiak, L.; Tryjanowski, P. Patterns in the distribution and directional asymmetry of fleas living on the northern white-breasted hedgehog *Erinaceus roumanicus*. *Folia Parasit.* **2017**, *64*, 026. [CrossRef] [PubMed]
36. Marchiondo, A.A.; Holdsworth, P.A.; Fourie, L.J.; Rugg, D.; Hellmann, K.; Snyder, D.E.; Dryden, M.W. World Association for the Advancement of Veterinary Parasitology (W.A.A.V.P.) second edition: Guidelines for evaluating the efficacy of parasiticides for the treatment; prevention and control of flea and tick infestations on dogs and cats. *Vet. Parasitol.* **2013**, *194*, 84–97. [PubMed]
37. Lewis, R.E. Notes on the geographical distribution and host preferences in the order Siphonaptera. Part 8. New taxa described between 1984 and 1990, with a current classification of the order. *J. Med. Entomol.* **1993**, *30*, 239–256. [CrossRef] [PubMed]
38. Hopkins, G.H.E.; Rothschild, M. *An Illustrated Catalogue of the Rothschild Collection of Fleas in the British Museum (Nat. Hist.), Vol. III. Hystrichopsyllidae*; Cambridge University Press: Cambridge, UK, 1962.
39. Beaucournu, J.C.; Launay, H. *Les Puces (Siphonaptera) de France et du Bassin Méditerranéen Occidental, Faune de France*; Fedération Française des Sociétés des Sciences Naturelles: Paris, France, 1990; Volume 76.
40. Lawrence, A.L.; Brown, G.K.; Peters, B.; Spielman, D.S.; Morin-Adeline, V.; Slapeta, J. High phylogenetic diversity of the cat flea (*Ctenocephalides felis*) at two mitochondrial DNA markers. *Med. Vet. Entomol.* **2014**, *28*, 330–336. [CrossRef] [PubMed]
41. Zurita, A.; Callejón, R.; de Rojas, M.; Cutillas, C. Morphological, biometrical and molecular characterization of *Archaeopsylla erinacei* (Bouché; 1835). *Bull. Entomol. Res.* **2018**, *108*, 726–738. [CrossRef]
42. van der Mescht, L.; Matthee, S.; Matthee, C.A. New taxonomic and evolutionary insights relevant to the cat flea, *Ctenocephalides felis*: A geographic perspective. *Mol. Phylogenet. Evol.* **2021**, *155*, 106990. [CrossRef]
43. Rohlf, F.J.; Marcus, L.F. A revolution morphometrics. *Trends Ecol. Evol.* **1993**, *8*, 129–132. [CrossRef]
44. Bookstein, F.L. Size and shape: A comment on semantics. *Syst. Zool.* **1989**, *38*, 173–180. [CrossRef]
45. Dujardin, J.P. Morphometrics applied to medical entomology. *Infect. Genet. Evol.* **2008**, *8*, 875–890. [CrossRef] [PubMed]
46. Dos Reis, S.P.; Pessoa, L.M.; Strauss, R.E. Application of size-free canonical discriminant analysis to studies of geographic differentiation. *Braz. J. Genet.* **1990**, *13*, 509–520.
47. Dujardin, J.P.; Le Pont, F. Geographical variation of metric properties within the neotropical sandflies. *Infect. Genet. Evol.* **2004**, *4*, 353–359. [CrossRef] [PubMed]
48. Dujardin, J.P. *BAC Software*; Institut de Recherches pour le Développement (IRD): Paris, France, 2002; Available online: http://www.fsf.org/copyleft/gpl.html (accessed on 10 April 2024).
49. Hii, S.F.; Lawrence, A.L.; Cuttell, L.; Tynas, R.; Abd Rani, P.A.M.; Šlapeta, J.; Traub, R.J. Evidence for a specifc host-endosymbiont relationship between '*Rickettsia* sp. genotype RF2125' and *Ctenocephalides felis orientis* infesting dogs in India. *Parasit. Vectors* **2015**, *8*, 169. [CrossRef] [PubMed]
50. Ménier, K.; Beaucournu, J.C. Taxonomic study of the genus *Ctenocephalides* Stiles & Collins; 1930 (Insecta: Siphonaptera: Pulicidae) by using aedeagus characters. *J. Med. Entomol.* **1998**, *35*, 883–890. [PubMed]
51. Güvendi, M.; Can, H.; Köseoğlu, A.F.; Alak, S.E.; Kandemir, C.; Taskın, T.; Sürgeç, E.; Demir, S.; Değirmenci Döşkaya, A.; Karakavuk, M.; et al. Investigation of the genetic diversity and flea-borne pathogens in *Ctenocephalides felis* samples collected from goats in İzmir and Şanlıurfa provinces of Turkey. *Comp. Immunol. Microbiol. Infect. Dis.* **2022**, *90–91*, 101896.
52. Chotelersak, K.; Puttikamonkul, S.; Samung, Y.; Chaiphongpachara, T.; Dujardin, J.P.; Sumruayphol, S. *Ctenocephalides orientis* and *Ctenocephalides felis* in Thailand: Head geometry by species, sex and geography. *Med. Vet. Entomol.* **2024**, *38*, 179–188. [CrossRef]
53. Ruenchit, P. State-of-the-Art Techniques for Diagnosis of Medical Parasites and Arthropods. *Diagnostics* **2021**, *11*, 1545. [CrossRef]
54. Ghosh, N.; Silva, J.; Vazquez, A.; Das, A.B.; Smith, D.W. Use of fluorescence and scanning electron microscopy as tools in teaching biology. In *Scanning Microscopies 2011: Advanced Microscopy Technologies for Defense, Homeland Security, Forensic, Life, Environmental, and Industrial Sciences*; Postek, M.T., Newbury, D.E., Platek, S.F., Joy, D.C., Maugel, T.K., Eds.; SPIE: Bellingham, WA, USA, 2011; Volume 8036, pp. 217–227.
55. García-Sánchez, A.M.; Rivero, J.; Callejón, R.; Zurita, A.; Reguera-Gomez, M.; Valero, M.A.; Cutillas, C. Differentiation of *Trichuris* species using a morphometric approach. *Int. J. Parasitol. Parasites Wildl.* **2019**, *31*, 218–223. [CrossRef]
56. Escobar-Chavarría, O.; Cossío-Bayúgar, R.; Ramírez-Noguera, P.; Prado-Ochoa, M.G.; Velázquez-Sánchez, A.M.; Muñoz-Guzmán, M.A.; Angeles, E.; Alba-Hurtado, F. In vivo and in vitro apoptosis induced by new acaricidal ethyl-carbamates in *Rhipicephalus microplus*. *Ticks Tick-Borne Dis.* **2021**, *12*, 101603. [CrossRef] [PubMed]

57. Kassa, B.; Lee, M.H.; Kumar, R.; Mickael, C.; Sanders, L.; Tuder, R.M.; Mentink-Kane, M.; Graham, B.B. Experimental *Schistosoma japonicum*-induced pulmonary hypertension. *PLoS Neglect. Trop. Dis.* **2022**, *16*, e0010343. [CrossRef] [PubMed]
58. Marrugal, A.; Callejón, R.; de Rojas, M.; Halajian, A.; Cutillas, C. Morphological, biometrical, and molecular characterization of *Ctenocephalides felis* and *Ctenocephalides canis* isolated from dogs from different geographical regions. *Parasitol. Res.* **2013**, *112*, 2289–2298. [CrossRef]

animals

MDPI

Article

New Insights on Tools for Detecting β-Tubulin Polymorphisms in *Trichuris trichiura* Using rhAmp^{TM} SNP Genotyping

Julia Rivero, Cristina Cutillas and Rocío Callejón *

Department of Microbiology and Parasitology, Faculty of Pharmacy, University of Seville, 41012 Seville, Spain; jrfernandez@us.es (J.R.); cutillas@us.es (C.C.)
* Correspondence: callejon@us.es; Tel.: +34-954566251

Simple Summary: Soil-transmitted worm infections, usually treated with benzimidazoles, can develop resistance due to genetic variations in a specific gene called β-tubulin isotype 1. This study aimed to create a new, quick, and accurate way to identify these genetic variations. We designed a test to spot changes in certain parts of the β-tubulin gene of *Trichuris trichiura*, the worm causing the infection. By using two different primers, we could distinguish between different genetic types at specific positions in the gene. We tested our method on samples from captive primates and found it to be reliable. Additionally, we explored whether the β-tubulin gene could be useful as a marker in genetic studies. Our tests worked well when we used them on samples from the field. However, we did not find any of the expected genetic variations in the worms or eggs we examined. Instead, all the samples showed the same genetic type. Despite this, our analysis of the β-tubulin gene confirmed the close relationship between *T. trichiura* and a related *Trichuris suis* species, which suggests that this gene could be valuable for understanding their evolutionary history.

Abstract: Soil-transmitted helminth (STH) infections, commonly treated with benzimidazoles, are linked to resistance through single nucleotide polymorphisms (SNPs) at position 167, 198, or 200 in the β-tubulin isotype 1 gene. The aim of this study was to establish a novel genotyping assay characterized by its rapidity and specificity. This assay was designed to detect the presence of SNPs within the partial β-tubulin gene of *Trichuris trichiura*. This was achieved through the biallelic discrimination at codons 167, 198, and 200 by employing the competitive binding of two allele-specific forward primers. The specificity and reliability of this assay were subsequently confirmed using *Trichuris* samples isolated from captive primates. Furthermore, a molecular study was conducted to substantiate the utility of the β-tubulin gene as a molecular marker. The assays showed high sensitivity and specificity when applied to field samples. Nevertheless, none of the SNPs within the β-tubulin gene were detected in any of the adult worms or eggs from the analyzed populations. All specimens consistently displayed an SS genotype. The examination of the β-tubulin gene further validated the established close relationships between the *T. trichiura* clade and *Trichuris suis* clade. This reaffirms its utility as a marker for phylogenetic analysis.

Keywords: anthelmintic resistance; *Trichuris trichiura*; β-tubulin gene; rhAmp^{TM} SNP genotyping; phylogeny

1. Introduction

Worldwide, soil-transmitted helminth (STH) infections are among the most common infections that can cause serious harm to human health. It is estimated 1.5 billion people, accounting 24% of the global population, are infected with these parasites, with a higher prevalence observed among preschool and school-age children. The main STH species are *Ascaris lumbricoides* (the roundworm), *Trichuris trichiura* (the whipworm), and *Necator americanus* and *Ancylostoma duodenale* (hookworms). These infections are transmitted through eggs found in human feces, which contaminate soil in regions with inadequate sanitation. This mainly affects impoverished and marginalized communities in tropical and subtropical areas, where the access to clean water, sanitation, and hygiene is limited [1].

The World Health Organization (WHO) has developed a strategy to control the infection of STHs. This strategy aims to regulate morbidity through the periodic treatment, known as preventive chemotherapy, of individuals at risk residing in endemic areas. The objective is to reduce and sustain low infection intensity and protect against morbidity by implementing large-scale mass drug administration programs. WHO recommends treatment with benzimidazoles (BZs) such as albendazoles and mebendazoles due to their effectiveness, affordability, and ease of administration by non-medical personnel [1]. BZs are known to exert their action by blocking the microtubule functions of parasites. This is achieved through the inhibition of β-tubulin polymerization in microtubules, leading to the subsequent inhibition of glucose uptake and transport. As a result, the parasites experience a deficiency of glycogen [2]. However, several studies suggest that the therapeutic efficacy of BZ against trichuriasis is progressively diminishing over time. This decline is believed to be partially attributed to the emergence of anthelminthic drug resistance (AR) [3–5], which develops because of prolonged and extensive reliance on BZ anthelmintics over an extended period of time [6]. Furthermore, the drugs are administered in single doses, and while this approach is operationally practical, it does not achieve 100% efficacy [4,7–9]. Therefore, this practice of administering suboptimal doses extensively over a prolonged period may contribute to the selection and development of AR. Additionally, periodic treatment has the potential to select for subpopulations of parasites that are resistant to the drugs [10–12]. Moreover, there are only a limited number of anthelmintic drugs that have been approved for the treatment of STH infections in humans [13,14].

Single nucleotide polymorphisms (SNPs) have been extensively employed for gene identification. Achieving allelic discrimination for a single SNP with a high degree of reliability and flexibility is of utmost importance for the precise detection of advantageous genes associated with specific SNP sites [15]. BZ resistance in *T. trichiura* is attributed to SNPs in the β-tubulin isotype 1 gene, specifically at codon 167, codon 200 (TTC > TAC), or codon 198 (GAG > GCG) [16–20]. In addition, the frequency of SNPs at codon 200 and 198 was found to increase following treatment, and it was significantly higher in individuals who exhibited a poor response to BZ compared to those who responded well [19]. To maintain the advantages of mass drug administration programs, it is crucial to have tools that can facilitate the large-scale detection of BZ resistance in human STHs. The current challenge of limited detection of phenotypic resistance may be attributed to multiple factors, including the absence of reliable and sensitive methods to monitor resistance genotypes before and after BZ treatment [21], a low frequency of resistance alleles, and the probability that BZ resistance is recessive, as in veterinary parasites [11].

To date, various platforms have been developed for genotyping individual SNPs. These include Kompetitive Allele Specific PCR (KASP) [22,23], RNase H2 enzyme-based amplification (rhAmp) [24], TaqMan [25], and semi-thermal asymmetric reverse PCR (STARP) [26]. Likewise, PCR-based methods, such as real-time PCR (RT-PCR), pyrosequencing, and genotyping assays using the SmartAmp2 method, have been developed for the detection of putative BZ resistance SNPs in human STH [17,18,27,28].

The main objectives of this study were (i) to develop a new genotyping assay, rhAmp™ SNP genotyping, for the screening of β-tubulin SNPs in *T. trichiura*; (ii) to assess the presence of BZ resistance-associated SNPs at positions 167, 198, and 200 within the β-tubulin gene in various populations of *T. trichiura* obtained from non-human primates (These SNPs are likely associated with BZ resistance in *Trichuris* spp.); and (iii) to conduct a molecular investigation aiming to validate the β-tubulin gene as a molecular marker applicable across different *Trichuris* spp. This validation was carried out to infer phylogenetic relationships between different clades of *Trichuris* spp. and detect the emergence of AR, thus gaining insights into its distribution among distinct clades.

2. Materials and Methods

2.1. Ethics Statement

This study did not require the approval of an ethics committee. Whipworms and eggs were isolated from stool and cecum samples from various vertebrate animal hosts. These animals were housed in zoological gardens and slaughterhouses in Spain and maintained with good animal practices.

2.2. Collection Samples

In this study, we have formulated two distinct sections. The first section is based on genotyping assays, for which we utilized *T. trichiura* samples, both eggs and adults, which were collected from different primate hosts. The primate host species analyzed were the Barbary macaque (*Macaca sylvanus*) and patas monkey (*Erythrocebus patas*), from Zoo Castellar (Cádiz, Spain), the vervet monkey (*Chlorocebus aethiops*) from Selwo Aventura (Málaga, Spain), and the Guinea baboon (*Papio papio*) from Parque de la Naturaleza de Cabárceno (Cantabria, Spain) (Table 1). The second section is centered on a phylogenetic analysis, also utilizing the previously mentioned samples, in addition to adult samples from suids (*Sus scrofa domestica*) from slaughterhouses in Seville and Huelva (Spain) and porcupine (*Hystrix cristata*) from Bioparc Fuengirola in Malaga, Spain (Table 1).

Table 1. Sequences of *Trichuris* spp. species obtained in the present study based on β-tubulin partial gene including sample ID, host species, geographical origin, GenBank accession numbers, length, and G + C content.

Species	Sample ID	Adult/Eggs	Host Species/Geographical Origin	Accession Number	Length (bp)	G + C Content (%)
Trichuris trichiura	TMSM9	Adult	*Macaca sylvanus*/Spain	OQ446467	445	44.94
Trichuris sp.	TMSF10	Adult	*Macaca sylvanus*/Spain	OQ446468	445	44.94
Trichuris trichiura	TPPM1	Adult	*Papio papio*/Spain	OQ446469	445	45.39
Trichuris trichiura	TPPF1	Adult	*Papio papio*/Spain	OQ446470	445	45.39
Trichuris trichiura	TPPF2	Adult	*Papio papio*/Spain	OQ446471	445	45.39
Trichuris trichiura	TCAE_1	Eggs	*Chlorocebus aethiops*/Spain	OQ446472	445	44.94
Trichuris trichiura	TCAE_2	Eggs	*Chlorocebus aethiops*/Spain	OQ446473	445	44.94
Trichuris trichiura	TCAE_3	Eggs	*Chlorocebus aethiops*/Spain	OQ446474	445	44.94
Trichuris trichiura	TCAE_4	Eggs	*Chlorocebus aethiops*/Spain	OQ446475	445	44.94
Trichuris trichiura	TEPE	Eggs	*Erythrocebus patas*/Spain	OQ446476	445	44.94
Trichuris colobae	TCO_901	Adult	*Colobus guereza kikuyensis*/Spain	OQ446477	445	44.94
Trichuris suis	TSF1	Adult	*Sus scrofa domestica*/Spain	OQ446478	445	44.72
Trichuris suis	TSF2	Adult	*Sus scrofa domestica*/Spain	OQ446479	445	44.72
Trichuris sp.	THCF10	Adult	*Hystrix cristata*/Spain	OQ446480	448	47.99
Trichuris sp.	THCF15	Adult	*Hystrix cristata*/Spain	OQ446481	448	47.99

Specimens were isolated from stool samples after treatment or collected from the cecum post-mortem and, consequently, washed in saline solution (0.9% *w*/*v*), and separately frozen at −20 °C until further analysis.

Sheather's sugar solution was used for egg concentration [29], and then, they were embryonated at 32 °C for 3 to 4 weeks with potassium dichromate solution (0.2% *w*/*v*) to provide moisture to the medium and to prevent the growth of fungi and bacteria [30].

2.3. Molecular Analysis

2.3.1. DNA Extraction

Whipworm identification and morpho-biometric analysis were performed in previous studies [31,32]. According to the manufacturer's protocol, total genomic DNA from samples (adult worms and batches of eggs) were extracted using DNeasy Blood and Tissue Kit (Qiagen, Hilden, Germany). To assess the quality of the extractions, 0.8% agarose gel electrophoresis infused with SYBR™ Safe DNA gel stained with 2% *w*/*v* Tris-Borate-EDTA (TBE) was used.

2.3.2. Genotyping by rhAmp SNP Assays

In this study, to carry out the genotyping analysis, we used a dual enzyme chemistry technology called rhAmp™ SNP genotyping [33]. This technology is based on the RNase H2-dependent polymerase chain reaction (rhPCR) and universal reporters [24]. Three SNPs from β-tubulin partial gene (codon 167, 198, and 200) were selected for the rhAmp assays. The selected SNPs with corresponding flanking sequences were submitted to the rhAmp™ Genotyping Design Tool at IDT (Integrate DNATechnologies, IDT; https://eu.idtdna.com/site/order/designtool/index/GENOTYPING_PREDESIGN (accessed on 12 February 2024)) and based on the strength of the thermodynamics, the highest ranked assays were retained. For primer design, flanking sequences shorter than 50 bp (base pairs) were extended up to 50–60 bp based on the *T. trichiura* reference β-tubulin sequence at the probe target sites to meet the technical requirements (Table 2). For each assay, rhAmp used two allele-specific primers and a locus-specific primer. For each SNP, synthetic gBlocks™ Gene fragments were used as known genotype controls during assays, where one represented the wild type (WT) and the other, the mutant allele (MA), and furthermore, both were mixed in an equal molar ratio to represent the heterozygous genotype (Table 3).

Table 2. Primers designed for rhAmp SNPs Assays for each codon using rhAmp™ Genotyping Design Tool at IDT.

Codon	*T. trichiura* rhAmp SNP Assays		
	Primer Type	**Primer**	**Sequence**
Codon 167	Allele-Specific	Primer 1	CCTGACCGAATTATGACAACTTT
	Allele-Specific	Primer 2	CCTGACCGAATTATGACAACTTA
	Locus-Specific	Primer	GCCCCAACGTGAACAGTATCAAA
Codon 198	Allele-Specific	Primer 1	GCTTCATTATCTATGCAGAATGTTT
	Allele-Specific	Primer 2	GCTTCATTATCTATGCAGAATGTTG
	Locus-Specific	Primer	GCATGCAACTCTGTCAGTCCA
Codon 200	Allele-Specific	Primer 1	AGCGCTTCATTATCTATGCAGAA
	Allele-Specific	Primer 2	AGCGCTTCATTATCTATGCAGAT
	Locus-Specific	Primer	GCATGCAACTCTGTCAGTCCA

Table 3. gBlocks® designed for β-tubulin partial gene of *T. trichiura* genotyping corresponding to the three SNPs (codon 167, 198, and 200). The SNPs are indicated in red color and in bold in the sequences.

***T. trichiura* gBlocks® Gene Fragments**	**Sequence**
gBlocks® WT	GAATCGGAAAGCTGCGACTGCCTGCAAGGGTTCCAGTTGACTCATTCCCTCGGCGGCG GAACTGGGAGTGGAATGGGTACGCTTCTGATATCTAAAATTCGGGAAGAGTATCCTGACC GAATTATGACAACTT**T**TAGTGTCGTTCCGTCTCCGAAGGCAAGTTGTTTGATACTGTTCACG TCGTGAACTATCGCCTTTTTAGGTTTCAGATACAGTTGTAGAACCATATAATGCAACTCTGT CAGTCCACCAGTTGGTAGAGAACACGGACG**A**AACA**T**TCTGCATAGATAATGAAGCGCTTT ACGATATTTGTTTCCGAACTTTGAAGTTAACAACACCAACTTACGGAGACTTAAATCATT TGGTTTCGGCAACCATGTCTGGAGTAACGACATGCCTACGCTTTCCTGGTCAGTTGAAT-GCTGATTTGCGGAAGCTGGCAGTC

Table 3. *Cont.*

T. trichiura gBlocks® Gene Fragments	Sequence
gBlocks® 167	GAATCGGAAAGCTGCGACTGCCTGCAAGGGTTCCAGTTGACTCATTCCCTCGGCGGCGG AACTGGGAGTGGAATGGGTACGCTTCTGATATCTAAAATTCGGGAAGAGTATCCTGACCG AATTATGACAACTTATAGTGTCGTTCCGTCTCCGAAGGCAAGTTGTTTGATACTGTTCACGTC GTGAACTATCGCCTTTTTAGGTTTCAGATACAGTTGTAGAACCATATAATGCAACTCTGTC AGTCCACCAGTTGGTAGAGAACACGGACGAAACATTCTGCATAGATAATGAAGCGCTTTA CGATATTTGTTTCCGAACTTTGAAGTTAACAACACCAACTTACGGAGACTTAAATCATTTG GTTTCGGCAACCATGTCTGGAGTAACGACATGCCTACGCTTTCCTGGTCAGTTGAATGCTG ATTTGCGGAAGCTGGCAGTC
gBlocks® 198	GAATCGGAAAGCTGCGACTGCCTGCAAGGGTTCCAGTTGACTCATTCCCTCGGCGGCG GAACTGGGAGTGGAATGGGTACGCTTCTGATATCTAAAATTCGGGAAGAGTATCCTGACC GAATTATGACAACTTTTAGTGTCGTTCCGTCTCCGAAGGCAAGTTGTTTGATACTGTTCACG TCGTGAACTATCGCCTTTTTAGGTTTCAGATACAGTTGTAGAACCATATAATGCAACTCTGT CAGTCCACCAGTTGGTAGAGAACACGGACGCAACATTCTGCATAGATAATGAAGCGCTTT ACGATATTTGTTTCCGAACTTTGAAGTTAACAACACCAACTTACGGAGACTTAAATCATTT GGTTTCGGCAACCATGTCTGGAGTAACGACATGCCTACGCTTTCCTGGTCAGTTGAATGCT GATTTGCGGAAGCTGGCAGTC
gBlocks® 200	GAATCGGAAAGCTGCGACTGCCTGCAAGGGTTCCAGTTGACTCATTCCCTCGGCGGCGG AACTGGGAGTGGAATGGGTACGCTTCTGATATCTAAAATTCGGGAAGAGTATCCTGACCG AATTATGACAACTTTTAGTGTCGTTCCGTCTCCGAAGGCAAGTTGTTTGATACTGTTCACGT CGTGAACTATCGCCTTTTTAGGTTTCAGATACAGTTGTAGAACCATATAATGCAACTCTGTC AGTCCACCAGTTGGTAGAGAACACGGACGAAACAATCTGCATAGATAATGAAGCGCTTTA CGATATTTGTTTCCGAACTTTGAAGTTAACAACACCAACTTACGGAGACTTAAATCATTTGG TTTCGGCAACCATGTCTGGAGTAACGACATGCCTACGCTTTCCTGGTCAGTTGAATGCTGA TTTGCGGAAGCTGGCAGTC

SNP genotyping assays were performed using 0.25 µL of rhAmp SNP Assays (20X), 2.65 µL of combined rhAmp Genotyping Master Mix (2X) and rhAmp Reporter Mix (40X), 0.10 µL of nuclease-free water and 2 µL of sample DNA, and 2 µL of control template (gBlocks fragments controls) or 2 µL of nuclease-free water (for no-template control reactions). Reactions were run on the CFX Connect Real-Time PCR Detection System (Bio-Rad Laboratories, Hercules, CA, USA), and analyses were carried out using CFX Maestro Software version 2.3 (Bio-Rad, Hercules, CA, USA). The thermal conditions were 95 °C for 10 min, followed by 40 cycles at 95 °C for 10 s, 60 °C for 30 s, and 68 °C for 20 s per the published protocol (www.idtdna.com/rhAmp-SNP-protocol (accessed on 29 February 2024).

The bi-allelic specificity of the rhAmp assays was provided by two probes, one labelled with Amidite-fluorescein (FAM™) dye and the other with Hexachloro-fluorescein (HEX™) dye (Table 4). These different dye reporters were independently detected with excitation sources and emission filters at the respective wavelengths. A total of 91 *Trichuris* genomic DNA samples were quantified. Hence, each specimen was called resistant (RR), susceptible (SS), and heterozygous (RS) in relation to the melting temperature obtained.

Table 4. Different excitation and emission spectra for fluorophores.

Dye	λ Excitation Filter (nm)	λ Emission Filter (nm)
FAM™	450	533
HEX™	483	568

2.3.3. PCR and Sequencing

In the samples analyzed in the present study, the partial molecular marker gene β-tubulin was amplified by a polymerase chain reaction (PCR) using a thermal cycler (Eppendorf AG and sequenced. (Hamburg, Germany). The primers and PCR conditions were previously described by Hansen et al. [34]. Amplification reactions consisted of 5 µL

(10 µM) of each primer, 5 µL of template DNA (50 ng/µL,) 25 µL of GoTaq G2 Green Master Mix, and nuclease-free water up to 50 µL. A negative (no-template DNA) control sample and a positive DNA control sample were included in each PCR reaction. PCR products were visualized on agarose gels (0.8%). Subsequently, bands were eluted and purified using the Wizard SV Gel and PCR Clean-Up System Kit (Promega, WI, USA). Once purified, PCR products were concentrated and sequenced in both directions by Stab Vida (Lisbon, Portugal).

2.3.4. Phylogenetic Studies

Accession numbers obtained in this study are available in the GenBank database (Table 1). To analyze the relationships among the different *Trichuris* species, additional sequences from GenBank database were included in the alignments (Supplementary Table S1).

The aligned nucleotide dataset was obtained by the MUSCLE alignment method in MEGA (Molecular Evolutionary Genetics Analysis) version 11 [35]. Moreover, the number of nucleotide differences per sequence was calculated to evaluate the identity among *Trichuris* β-tubulin partial sequences by Compute Pairwise Distances based on the number of differences method of MEGA11 [35].

All phylogenetic trees were inferred by two different methods, Maximum Likelihood (ML) and Bayesian Inferences (BIs). To generate the ML tree, PhyML 3.0 package [36] was used, and for the BI tree, MrBayes v3.2.6 [37] was used. To resolve the best-fit substitution model for the nucleotide dataset jModelTest [38] was employed, and the models of evolution were determined in agreement with Akaike Information Criterion [36,39]. Bootstrapping (heuristic option) of more than 1000 replications was used to examine the topology support to assess the relative reliability of the clades, and the Bayesian Posterior Probabilities (BPPs) comprised the percentage converted. Standard deviation of split frequencies was used to determine if the number of generations completed was adequate. Each dataset was run for 10 million generations, and the chain was sampled every 500 generations. In addition, trees from the first million generations were discarded based on an assessment of convergence. Empirically, during the burn, the examination of the log-likelihood values of the chain was carried out.

3. Results

3.1. SNP Genotyping Assays

rhAmp^{TM} SNP Genotyping were completely optimized at codon 167, 198, and 200 of the β-tubulin gene in *T. trichiura* and were able to detect the presence or absence of the WT and MA genotypes.

Before real-time analysis, for each SNP analyzed, standard peaks were created for RR (homozygote-resistant), SS (homozygote-susceptible), and RS (heterozygote) using WT, MA, and heterozygote DNA samples obtained commercially from IDT.

For each point mutation site, the melting peaks obtained from RT-PCR were analyzed and determined as RR, SS, and RS as reported by the specific Tm. Ninety-one Trichuris samples were collected from non-human primates, and genotyping assays by RT-PCR revealed the different dots, homozygote and heterozygote, obtained in the present work (Figure 1). Positive (WT, MA, and heterozygote) and negative controls were always included, and no amplification in negative controls was observed.

Out of the three analyzed SNPs in the partial β-tubulin gene, all specimens exhibited SS (100%). However, for codon 168, low signal intensity was observed in all twenty-one egg batch samples for both fluorophores. For codon 198, only four egg batch samples showed low signal intensity, and for codon 200, seven egg batch samples displayed low signal intensity. To ensure the accuracy of the results, all experiments were repeated twice. In concordance with these results, the resistance allele frequency (RAF) was 0.0 for each SNP studied.

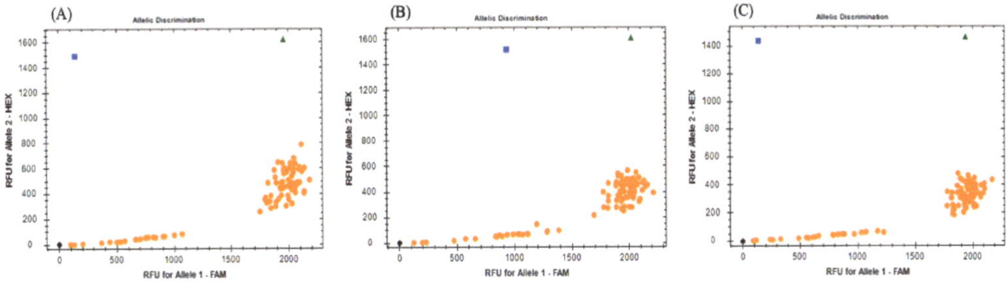

Figure 1. Allelic discrimination plots obtained for rhAmp^{TM} SNP genotyping assays targeting the three SNPs: (**A**) codon 167; (**B**) codon 198; (**C**) codon 200. Homozygous genotypes are represented by orange dots and blue squares, where the orange dots correspond to an SS and the blue squares to an RR, the green triangles represent heterozygous genotypes, and the black rhombus on the bottom left of the plot are no-template controls.

3.2. Molecular Analysis

The phylogenetic tree inferred by ML and BI methods revealed three main clades (Figure 2). Clade 1 consisted of sequences from both, *Trichuris* sp. and *T. trichiura*, isolated from humans and non-human primates (100 BPP and 94% ML BV). In the BI analysis, the sequences were in polytomy, but the support for different subclades was not strong based on ML methods. Clade 2 consisted of sequences from *Trichuris colobae* and *Trichuris suis* (100 BPP and 89% ML BV), which were grouped into two distinct and well-supported subclades. Further, clade 3 consisted of sequences from *Trichuris* sp. isolated from *H. cristata* that were strongly supported (100 BPP and 85% ML BV) (Figure 2).

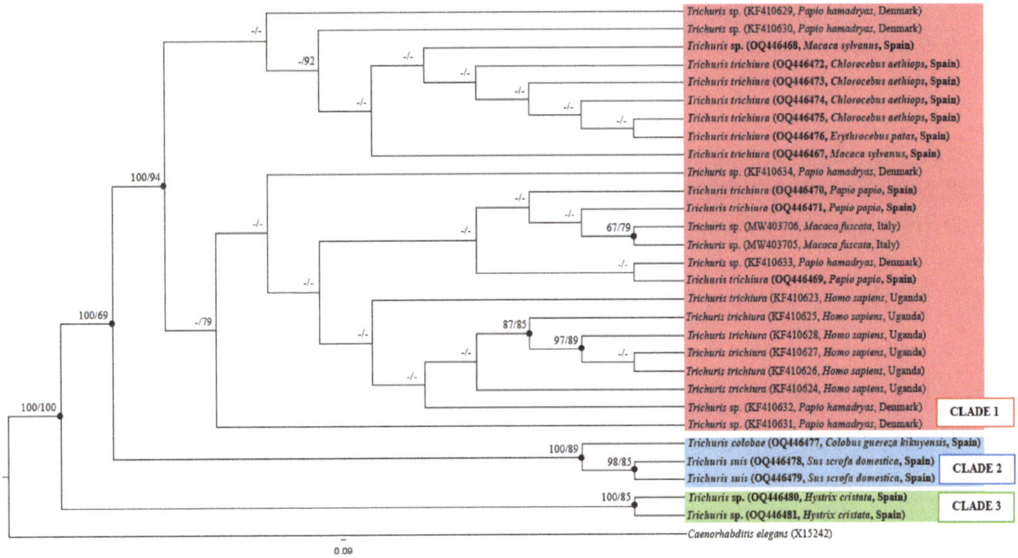

Figure 2. Phylogenetic tree inferred based on the molecular marker β-tubulin partial gene of *Trichuris* species using ML method. Bayesian Posterior Probabilities (BPPs) are listed first, followed by ML bootstrap values (BVs) of clades, for clade frequencies exceeding 60%.

The results obtained were consistent with the identity analysis, where sequences clustered within the same clade displayed an inter-populational similarity percentage of over 98.7% for clade 1, ranging from 98 to 100% for clade 2, and 100% for clade 3 (Supplementary Table S2a,b).

The similarity percentage between clade 1 and clade 2 ranged from 91.3 to 92.3%, between clade 1 and clade 3, 81.8–82.9%, and between clade 2 and clade 3, 82.9–83.1% (Supplementary Table S2a,b).

Phylogenetic analysis revealed a sister relationship between clade 1, which included *T. trichiura* and *Trichuris* sp. isolated from humans and non-human primates, and clade 2, which comprised *T. suis* and *T. colobae* (100 BPP and 69% ML BV). Likewise, both clades remained separate from *Trichuris* sp. from porcupine (Figure 2).

4. Discussion

STHs have been widely treated with mass drug administration, a highly effective approach in reducing helminth-related morbidity by limiting transmission within endemic communities. However, while this strategy brings numerous advantages to the population, it may also lead to unintended consequences, such as the gradual reduction in treatment effectiveness [40,41]. Consequently, with the expansion of drug donations, the emergence of AR becomes increasingly probable. This decline in treatment efficacy has been particularly concerning in veterinary parasites, where certain nematode species have demonstrated high levels of resistance to drugs, including BZ [11,12,42–44]. Moreover, in these helminths, SNPs in the β-tubulin isotype 1 gene have been associated with resistance to BZ [10–12].

Several tests have been proposed to identify mutations related to BZ resistance in helminths [19,45,46]. The SmartAmp2 constituted an alternative approach utilized. This method streamlines the process by implementing a single-step protocol, enabling enhanced expediency, showcasing distinct advantages over conventional PCR. Additionally, it has demonstrated efficacy in the investigation of diverse STHs identified in fecal samples. Nevertheless, utilizing this technique, the optimization of all necessary SNP detections proved unattainable, as the identification of codon 167 in *T. trichiura* was unsuccessful [28]. Diarra et al. [19] conducted an evaluation of the SNPs in codon 200 in A. lumbricoides using pyrosequencing, which necessitates specialized equipment. Another method employed was the RFLP-PCR, which is considered simpler and more sensitive but has certain limitations, such as cases where the DNA sequences are not recognized by commercially available restriction enzymes or possess multiple recognition sites for a single enzyme [45]. In contrast, Furtado et al. [46] employed ARMS-PCR, a technique that solely requires a conventional thermocycler, offering a straightforward and immediate result. Broccanello et al. [47] conducted a comparison of the accuracy, sensitivity, and costs of TaqMan, KASP, and rhAmp^{TM} SNP Genotyping methods in sugar beet (*Beta vulgaris* L.). The sensitivity test revealed that both TaqMan and rhAmp were able to accurately determine SNP genotypes. In the case of rhAmp^{TM} SNP Genotyping, 24 of the 33 SNPs exhibited 100% concordance with other two technologies. The genotype concordance with both technologies for the remaining nine targets exceeded 99%.

This study represents the initial endeavor to establish rhAmp assays for genotyping SNPs in the β-tubulin partial gene of *Trichuris* spp. samples. The primary objective was to develop a rapid and highly sensitive method capable of discriminating between different alleles in *Trichuris* samples, while also determining the genotyping frequencies. The developed rhAmp methods proved effective in accurately identifying the three different genotypes. Sequencing of the PCR products further confirmed the high conservation of the region flanked by the primers (PCR) and probes (rhAmp) which also contained the mutation.

None of the analyzed SNPs were found in our examined samples. Hence, the prevalence observed in our study was found to be lower compared to previously reported prevalence rates for other parasites. For instance, the prevalence of A. lumbricoides was reported to be 0.5% [46], while *Haemonchus contortus* showed a much higher prevalence of 74% [48]. Nonetheless, certain studies examining theses SNPs in parasites of veterinary significance, such as *P. equorum* and *Ascaridia galli*, have failed to detect these alterations, even in parasite populations subjected to frequent treatments throughout the year [49–51]. However, our findings are consistent with those obtained by Hansen et al. [34], wherein none of the 27 adult worms and 39 egg batched of *Trichuris* from human analyzed, or the 49 adult worms of *Trichuris* samples from baboons, exhibited mutations in any of the analyzed SNPs. This suggests that the identified SNPs may not be the primary mechanism responsible for BZ resistance in these nematode species.

The methodology proposed here offers a robust tool for screening the emergence of anthelmintic resistance mutations in populations of parasitic nematodes. Furthermore, it represents a rapid, highly sensitive, and specific technique that obviates the need for PCR amplification, a thermocycler, and electrophoresis.

Additionally, according to previous phylogenetic studies carried out by other authors, which were based on both nuclear and mitochondrial markers and focused on *Trichuris* species, two main clades have been identified. These clades have been previously referred to as the "*T. trichiura* lineage", which parasitizes humans and non-human primates, and the "*T. suis* lineage", which infects suids and other primates, such as *Colobus guereza kikuyensis* or *Papio ursinus* [31,32]. Furthermore, various authors have reported the hypothesis that a complex whipworm species may exist in primates, suggesting that different *Trichuris* species infect both primates and humans [32,52–56]. In addition, our results align with those obtained by Rivero et al. [57], where, based on nuclear and mitochondrial markers as well, the *Trichuris* sequences isolated from porcupines were distinct from the rest of the analyzed sequences, remaining isolated from the rest of the sequences in the obtained tree. As a result, the results obtained confirmed the sister relationship between clade 1 and clade 2 as identified in the phylogenetic tree. This finding corroborates the utility of β-tubulin as an additional marker, in addition to those previously described, for the phylogenetic analysis of the *Trichuris* species.

5. Conclusions

The current study provides a novel genotyping assay for assessing the prevalence of frequency of AR-associated β-tubulin SNPs at codons 167, 198, and 200 in *T. trichiura*. This investigation has showcased the efficacy of rhAmp methods in accurately discriminating and identifying β-tubulin SNPs within *T. trichiura* adults and eggs. Consequently, rhAmp^{TM} SNP Genotyping can be considered a molecular tool that is both rapid and sensitive for the in vitro assessment of BZ susceptibility in *Trichuris* spp. Moreover, using pooled samples, it can be proposed as a cost-effective method, offering immediate practical advantages in the field.

Furthermore, this study provides further insights into the phylogeny of β-tubulin in *Trichuris* spp., reaffirming the previously established close relationship between the *T. trichiura* clade and *T. suis* clade. Thus, we affirm its usefulness as a marker for phylogenetic analysis.

Supplementary Materials: The following supporting information can be downloaded at: https://www.mdpi.com/article/10.3390/ani14111545/s1, Table S1: Sequences of *Trichuris* spp. and outgroups species obtained from GenBank and used for phylogenetic analysis; Table S2a: Intra-specific and inter-specific similarity observed in β-tubulin partial gene sequences in *Trichuris* species isolated from different hosts. Values are given in percentages (%); Table S2b: Continuation of Table S2a. Intra-specific and inter-specific similarity observed in β-tubulin partial gene sequences in *Trichuris* species isolated from different hosts. Values are given in percentages (%).

Author Contributions: Conceptualization, R.C. and J.R.; methodology, J.R. and R.C.; validation, J.R. and R.C.; formal analysis, J.R.; investigation, J.R. and R.C.; data curation, J.R.; writing—original draft preparation, J.R.; writing—review and editing, J.R., R.C. and C.C.; supervision, C.C.; project administration, C.C. All authors have read and agreed to the published version of the manuscript.

Funding: This research received no external funding.

Institutional Review Board Statement: Not applicable.

Informed Consent Statement: Not applicable.

Data Availability Statement: The data presented in this study are available on request from the corresponding author.

Acknowledgments: The authors want to extend their gratitude to the zoological gardens (Zoo Castellar, Selwo Aventura, Parque Naturaleza de Cabárceno and Bioparc Fuengirola) for providing the analyzed samples.

Conflicts of Interest: The authors declare no conflicts of interest.

References

1. World Health Organization. Soil-Transmitted Infections. Available online: https://www.who.int/news-room/fact-sheets/detail/soil-transmitted-helminth-infections (accessed on 18 January 2023).
2. Kern, P. *Echinococcus granulosus* infection: Clinical presentation, medical treatment and outcome. *Langenbecks Arch. Surg.* **2003**, *388*, 413–420. [CrossRef] [PubMed]
3. Soukhathammavong, P.A.; Sayasone, S.; Phongluxa, K.; Xayaseng, V.; Utzinger, J.; Vounatsou, P.; Hatz, C.; Akkhavong, K.; Keiser, J.; Odermatt, P. Low efficacy of single-dose albendazole and mebendazole against hookworm and effect on concomitant helminth infection in Lao PDR. *PLoS Negl. Trop. Dis.* **2012**, *6*, e1417. [CrossRef] [PubMed]
4. Moser, W.; Schindler, C.; Keiser, J. Efficacy of recommended drugs against soil transmitted helminths: Systematic review and network meta-analysis. *BMJ* **2017**, *358*, j4307. [CrossRef] [PubMed]
5. Moser, W.; Schindler, C.; Keiser, J. Drug Combinations Against Soil-Transmitted Helminth Infections. *Adv. Parasitol.* **2019**, *103*, 91–115. [CrossRef] [PubMed]
6. Kaplan, R.M. Drug resistance in nematodes of veterinary importance: A status report. *Trends Parasitol.* **2004**, *20*, 477–481. [CrossRef] [PubMed]
7. Keiser, J.; Utzinger, J. Efficacy of current drugs against soil-transmitted helminth infections: Systematic review and meta-analysis. *JAMA* **2008**, *299*, 1937–1948. [CrossRef] [PubMed]
8. Vercruysse, J.; Behnke, J.M.; Albonico, M.; Ame, S.M.; Angebault, C.; Bethony, J.M.; Engels, D.; Guillard, B.; Nguyen, T.V.; Kang, G.; et al. Assessment of the anthelmintic efficacy of albendazole in school children in seven countries where soil-transmitted helminths are endemic. *PLoS Negl. Trop. Dis.* **2011**, *5*, e948. [CrossRef] [PubMed]
9. Levecke, B.; Montresor, A.; Albonico, M.; Ame, S.M.; Behnke, J.M.; Bethony, J.M.; Noumedem, C.D.; Engels, D.; Guillard, B.; Kotze, A.C.; et al. Assessment of anthelmintic efficacy of mebendazole in school children in six countries where soil-transmitted helminths are endemic. *PLoS Negl. Trop. Dis.* **2014**, *8*, e3204. [CrossRef]
10. Kwa, M.S.; Veenstra, J.G.; Roos, M.H. Benzimidazole resistance in *Haemonchus contortus* is correlated with a conserved mutation at amino acid 200 in beta-tubulin isotype 1. *Mol. Biochem. Parasitol.* **1994**, *63*, 299–303. [CrossRef] [PubMed]
11. Prichard, R. Genetic variability following selection of *Haemonchus contortus* with anthelmintics. *Trends Parasitol.* **2001**, *17*, 445–453. [CrossRef]
12. Ghisi, M.; Kaminsky, R.; Maser, P. Phenotyping and genotyping of *Haemonchus contortus* isolates reveals a new putative candidate mutation for benzimidazole resistance in nematodes. *Vet. Parasitol.* **2007**, *144*, 313–320. [CrossRef] [PubMed]
13. Keiser, J.; Utzinger, J. The drugs we have and the drugs we need against major helminth infections. *Adv. Parasitol.* **2010**, *73*, 197–230. [CrossRef] [PubMed]
14. Olliaro, P.; Seiler, J.; Kuesel, A.; Horton, J.; Clark, J.N.; Don, R.; Keiser, J. Potential drug development candidates for human soil-transmitted helminthiases. *PLoS Negl. Trop. Dis.* **2011**, *5*, e1138. [CrossRef] [PubMed]
15. Tehseen, M.M.; Zheng, Y.; Wyatt, N.A.; Bolton, M.D.; Yang, S.; Xu, S.S.; Li, X.; Chu, C. Development of STARP Marker Platform for Flexible SNP Genotyping in Sugarbeet. *Agronomy* **2023**, *13*, 1359. [CrossRef]
16. Bennett, A.B.; Anderson, T.J.; Barker, G.C.; Michael, E.; Bundy, D.A. Sequence variation in the *Trichuris trichiura* beta-tubulin locus: Implications for the development of benzimidazole resistance. *Int. J. Parasitol.* **2002**, *32*, 1519–1528. [CrossRef] [PubMed]
17. Diawara, A.; Drake, L.J.; Suswillo, R.R.; Kihara, J.; Bundy, D.A.; Scott, M.E.; Halpenny, C.; Stothard, J.R.; Prichard, R.K. Assays to detect beta-tubulin codon 200 polymorphism in *Trichuris trichiura* and *Ascaris lumbricoides*. *PLoS Negl. Trop. Dis.* **2009**, *3*, e397. [CrossRef] [PubMed]
18. Diawara, A.; Schwenkenbecher, J.M.; Kaplan, R.M.; Prichard, R.K. Molecular and biological diagnostic tests for monitoring benzimidazole resistance in human soil-transmitted helminths. *Am. J. Trop. Med. Hyg.* **2013**, *88*, 1052–1061. [CrossRef]
19. Diawara, A.; Halpenny, C.M.; Churcher, T.S.; Mwandawiro, C.; Kihara, J.; Kaplan, R.M.; Streit, T.G.; Idaghdour, Y.; Scott, M.E.; Basáñez, M.G.; et al. Association between response to albendazole treatment and β-tubulin genotype frequencies in soil-transmitted helminths. *PLoS Negl. Trop. Dis.* **2013**, *30*, *7*:e2247. [CrossRef] [PubMed]
20. Mendes de Oliveira, V.N.G.; Zuccherato, L.W.; Dos Santos, T.R.; Rabelo, É.M.L.; Furtado, L.F.V. Detection of Benzimidazole Resistance-Associated Single-Nucleotide Polymorphisms in the Beta-Tubulin Gene in *Trichuris trichiura* from Brazilian Populations. *Am. J. Trop. Med. Hyg.* **2022**, *107*, 640–648. [CrossRef]
21. Vercruysse, J.; Levecke, B.; Prichard, R. Human soil-transmitted helminths: Implications of mass drug administration. *Curr. Opin. Infect. Dis.* **2012**, *25*, 703–708. [CrossRef]
22. Semagn, K.; Babu, R.; Hearne, S.; Olsen, M. Single nucleotide polymorphism genotyping using Kompetitive Allele Specific PCR (KASP): Overview of the technology and its application in crop improvement. *Mol. Breed.* **2014**, *33*, 1–14. [CrossRef]

23. Ertiro, B.T.; Ogugo, V.; Worku, M.; Das, B.; Olsen, M.; Labuschagne, M.; Semagn, K. Comparison of Kompetitive Allele Specific PCR (KASP) and genotyping by sequencing (GBS) for quality control analysis in maize. *BMC Genom.* **2015**, *16*, 908. [CrossRef] [PubMed]
24. Dobosy, J.R.; Rose, S.D.; Beltz, K.R.; Rupp, S.M.; Powers, K.M.; Behlke, M.A.; Walder, J.A. RNase H-dependent PCR (rhPCR): Improved specificity and single nucleotide polymorphism detection using blocked cleavable primers. *BMC Biotechnol.* **2011**, *11*, 80. [CrossRef] [PubMed]
25. Woodward, J. Bi-allelic SNP genotyping using the TaqMan assay. *Methods Mol. Biol.* **2014**, *1145*, 67–74. [CrossRef]
26. Long, Y.M.; Chao, W.S.; Ma, G.J.; Xu, S.S.; Qi, L.L. An innovative SNP genotyping method adapting to multiple platforms and throughputs. *Theor. Appl. Genet.* **2017**, *130*, 597–607. [CrossRef] [PubMed]
27. Schwenkenbecher, J.M.; Albonico, M.; Bickle, Q.; Kaplan, R.M. Characterization of beta-tubulin genes in hookworms and investigation of resistance-associated mutations using real-time PCR. *Mol. Biochem. Parasitol.* **2007**, *156*, 167–174. [CrossRef] [PubMed]
28. Rashwan, N.; Scott, M.; Prichard, R. Rapid Genotyping of β-tubulin Polymorphisms in *Trichuris trichiura* and *Ascaris lumbricoides*. *PLoS Negl. Trop. Dis.* **2017**, *11*, e0005205. [CrossRef] [PubMed]
29. Gupta, S.K.; Singla, L.D. Chapter 6: Diagnostic Trends in Parasitic Diseases of Animals. In *Veterinary Diagnostics*; Gupta, R.P., Garg, S.R., Nehra, V., Lather, D., Eds.; Statish Serial Publishing House: Delhi, India, 2013; pp. 81–112.
30. Horii, Y.; Usui, M. Experimental transmission of *Trichuris* ova from monkeys to man. *Trans. R. Soc. Trop. Med. Hyg.* **1985**, *79*, 423. [CrossRef] [PubMed]
31. Rivero, J.; García-Sánchez, Á.M.; Zurita, A.; Cutillas, C.; Callejón, R. *Trichuris trichiura* isolated from *Macaca sylvanus*: Morphological, biometrical, and molecular study. *BMC Vet. Res.* **2020**, *16*, 445, Erratum in: *BMC Vet. Res.* **2021**, *17*, 160. [CrossRef]
32. Rivero, J.; Cutillas, C.; Callejón, R. *Trichuris trichiura* (Linnaeus, 1771) From Human and Non-human Primates: Morphology, Biometry, Host Specificity, Molecular Characterization, and Phylogeny. *Front. Vet. Sci.* **2021**, *7*, 626120. [CrossRef]
33. Beltz, K.; Tsang, D.; Wang, J.; Rose, S.; Bao, Y.; Wang, Y.; Larkin, K.; Rupp, S.; Schrepfer, D.; Datta, K.; et al. A high-performing and cost-effective SNP genotyping method using rhPCR and universal reporters. *Adv. Biosci. Biotechnol.* **2018**, *9*, 497–512. [CrossRef]
34. Hansen, T.V.; Thamsborg, S.M.; Olsen, A.; Prichard, R.K.; Nejsum, P. Genetic variations in the beta-tubulin gene and the internal transcribed spacer 2 region of *Trichuris* species from man and baboons. *Parasit. Vectors* **2013**, *6*, 236. [CrossRef] [PubMed]
35. Tamura, K.; Stecher, G.; Kumar, S. MEGA11: Molecular Evolutionary Genetics Analysis version 11. *Mol. Biol. Evol.* **2021**, *38*, 3022–3027. [CrossRef] [PubMed]
36. Guindon, S.; Dufayard, J.F.; Lefort, V.; Anisimova, M.; Hordijk, W.; Gascuel, O. New algorithms and methods to estimate maximum-likelihood phylogenies: Assessing the performance of PhyML 3.0. *Syst. Biol.* **2010**, *59*, 307–321. [CrossRef] [PubMed]
37. Ronquist, F.; Huelsenbeck, J.P. MrBayes 3: Bayesian phylogenetic inference under mixed models. *Bioinformatics* **2003**, *19*, 1572–1574. [CrossRef]
38. Posada, D. jModelTest: Phylogenetic model averaging. *Mol. Biol. Evol.* **2008**, *25*, 1253–1256. [CrossRef] [PubMed]
39. Posada, D.; Buckley, T.R. Model selection and model averaging in phylogenetics: Advantages of akaike information criterion and bayesian approaches over likelihood ratio tests. *Syst. Biol.* **2004**, *53*, 793–808. [CrossRef] [PubMed]
40. Humphries, D.; Nguyen, S.; Boakye, D.; Wilson, M.; Cappello, M. The promise and pitfalls of mass drug administration to control intestinal helminth infections. *Curr. Opin. Infect. Dis.* **2012**, *25*, 584–589. [CrossRef] [PubMed]
41. Furtado, L.F.; de Paiva Bello, A.C.P.; Rabelo, É.M.L. Benzimidazole resistance in helminths: From problem to diagnosis. *Acta Trop.* **2016**, *162*, 95–102. [CrossRef]
42. Geerts, S.; Gryseels, B. Anthelmintic resistance in human helminths: A review. *Trop. Med. Int. Health* **2001**, *6*, 915–921. [CrossRef]
43. van Wyk, J.A. Refugia-overlooked as perhaps the most potent factor concerning the development of anthelmintic resistance. *J. Vet. Res.* **2001**, *68*, 55–67.
44. Wolstenholme, A.J.; Fairweather, I.; Prichard, R.; von Samson-Himmelstjerna, G.; Sangster, N.C. Drug resistance in veterinary helminths. *Trends Parasitol.* **2004**, *20*, 469–476. [CrossRef]
45. Furtado, L.F.V.; Magalhães, J.G.S.; Rabelo, É.M.L. Standardization and application of a modified RFLP-PCR methodology for analysis of polymorphisms linked to treatment resistance in *Ancylostoma braziliense*. *Parasit. Vectors* **2018**, *11*, 540. [CrossRef]
46. Furtado, L.F.V.; Medeiros, C.d.S.; Zuccherato, L.W.; Alves, W.P.; de Oliveira, V.N.G.M.; da Silva, V.J.; Miranda, G.S.; Fujiwara, R.T.; Rabelo, É.M.L. First identification of the benzimidazole resistance-associated F200Y SNP in the beta-tubulin gene in *Ascaris lumbricoides*. *PLoS ONE* **2019**, *14*, e0224108. [CrossRef] [PubMed]
47. Broccanello, C.; Chiodi, C.; Funk, A.; McGrath, J.M.; Panella, L.; Stevanato, P. Comparison of three PCR-based assays for SNP genotyping in plants. *Plant Methods* **2018**, *14*, 1–8. [CrossRef] [PubMed]
48. Dixit, A.K.; Das, G.; Dixit, P.; Singh, A.P.; Kumbhakar, N.K.; Sankar, M.; Sharma, R.L. An assessment of benzimidazole resistance against caprine nematodes in Central India. *Trop. Anim. Health Prod.* **2017**, *49*, 1471–1478. [CrossRef] [PubMed]
49. Tydén, E.; Engström, A.; Morrison, D.A.; Höglund, J. Sequencing of the β-tubulin genes in the ascarid nematodes *Parascaris equorum* and *Ascaridia galli*. *Mol. Biochem. Parasitol.* **2013**, *190*, 38–43. [CrossRef] [PubMed]
50. Tydén, E.; Dahlberg, J.; Karlberg, O.; Höglund, J. Deep amplicon sequencing of preselected isolates of *Parascaris equorum* in β-tubulin codons associated with benzimidazole resistance in other nematodes. *Parasit. Vectors* **2014**, *7*, 410. [CrossRef]
51. Tarbiat, B.; Jansson, D.S.; Tydén, E.; Höglund, J. Evaluation of benzimidazole resistance status in *Ascaridia galli*. *Parasitology* **2017**, *144*, 1338–1345. [CrossRef]

52. Liu, G.H.; Gasser, R.B.; Su, A.; Nejsum, P.; Peng, L.; Lin, R.Q.; Li, M.W.; Xu, M.J.; Zhu, X.Q. Clear genetic distinctiveness between human-and pig-derived *Trichuris* based on analysis of mitochondrial datasets. *PLoS Negl. Trop. Dis.* **2012**, *6*, e1539. [CrossRef]
53. Cavallero, S.; De Liberato, C.; Friedrich, K.G.; Di Cave, D.; Masella, V.; D'Amelio, S.; Berrilli, F. Genetic heterogeneity and phylogeny of *Trichuris* spp. from captive non-human primates based on ribosomal DNA sequence data. *Infect. Genet. Evol.* **2015**, *34*, 450–456. [CrossRef] [PubMed]
54. Cavallero, S.; Nejsum, P.; Cutillas, C.; Callejón, R.; Doležalová, J.; Modrý, D.; D'Amelio, S. Insights into the molecular systematics of *Trichuris* infecting captive primates based on mitochondrial DNA analysis. *Vet. Parasitol.* **2019**, *272*, 23–30. [CrossRef] [PubMed]
55. Hawash, M.B.; Andersen, L.O.; Gasser, R.B.; Stensvold, C.; Nejsum, P. Mitochondrial genome analyses suggest multiple *Trichuris* species in humans, baboons, and pigs from different geographical regions. *PLoS Negl. Trop. Dis.* **2015**, *9*, e0004059. [CrossRef] [PubMed]
56. Xie, Y.; Zhao, B.; Hoberg, E.P.; Li, M.; Zhou, X.; Gu, X.; Lai, W.; Peng, X.; Yang, G. Genetic characterisation and phylogenetic status of whipworms (*Trichuris* spp.) from captive non-human primates in China, determined by nuclear and mitochondrial sequencing. *Parasit. Vectors* **2018**, *11*, 516. [CrossRef]
57. Rivero, J.; García-Sánchez, Á.M.; Callejón, R.; Cutillas, C. Characterization of *Trichuris* species from porcupine (*Hystrix cristata*) at zoological garden of Spain. *Acta Trop.* **2022**, *228*, 106276. [CrossRef]

animals

MDPI

Article

Gastrointestinal Parasites in Owned Dogs in Serbia: Prevalence and Risk Factors

Nemanja M. Jovanovic [1], Olga Bisenic [2], Katarina Nenadovic [3], Danica Bogunovic [1], Milan Rajkovic [1,*], Milan Maletic [4], Milorad Mirilovic [5] and Tamara Ilic [1]

1 Department of Parasitology, Faculty of Veterinary Medicine, University of Belgrade, Bul. Oslobodjenja 18, 11000 Belgrade, Serbia; nmjovanovic@vet.bg.ac.rs (N.M.J.); danicab@vet.bg.ac.rs (D.B.); tamara@vet.bg.ac.rs (T.I.)

2 Faculty of Veterinary Medicine, University of Belgrade, Bul. Oslobodjenja 18, 11000 Belgrade, Serbia; olga.bisenic@gmail.com

3 Department of Animal Hygiene, Faculty of Veterinary Medicine, University of Belgrade, Bul. Oslobodjenja 18, 11000 Belgrade, Serbia; katarinar@vet.bg.ac.rs

4 Department of Reproduction, Fertility and Artificial Insemination, Faculty of Veterinary Medicine, University of Belgrade, Bul. Oslobodjenja 18, 11000 Belgrade, Serbia; maletic@vet.bg.ac.rs

5 Department of Economics and Statistics, Faculty of Veterinary Medicine, University of Belgrade, Bul. Oslobodjenja 18, 11000 Belgrade, Serbia; mija@vet.bg.ac.rs

* Correspondence: mrajkovic@vet.bg.ac.rs

Simple Summary: This research conducted in Serbia aimed to identify intestinal parasites in dogs that could potentially infect humans. Total prevalence of intestinal endoparasites was 62.6%. Various endoparasites such as *Cystoisospora* spp., *Sarcocystis* spp., *Neospora caninum/Hammondia* spp., *Giardia intestinalis*, *Toxocara canis*, *Toxascaris leonina*, Ancylostomatidae, *Trichuris vulpis*, *Capillaria* spp., *Alaria alata* and Taeniidae were found. Factors like age, outdoor living, attitude and diet were linked to higher infection rates. This study emphasizes the importance of educating dog owners, conducting routine parasitological tests on their pets and regular deworming strategies.

Abstract: Dogs are the most popular pets worldwide. Close contact between dogs and people increases the risk of transmission of various zoonotic parasitic infections. Given the importance of veterinary medicine in preserving the One Health concept, the aim of this research was to identify intestinal parasites that may have zoonotic potential and to evaluate risk factors (individual and environmental). The research was conducted in Serbia in 2022 and 2023 on 382 owned dogs, using qualitative methods of coprological examination with a concentration on parasitic elements. The overall prevalence of intestinal parasites was 62.6%, with the following detected: protozoa: *Cystoisospora* spp. (9.2%), *Sarcocystis* spp. (4.5%), *Neospora caninum/Hammondia* spp. (3.7%), *Giardia intestinalis* (11.8%); nematoda: *Toxocara canis* (11.5%), *Toxascaris leonina* (4.2%), family Ancylostomatidae (38.0%), *Trichuris vulpis* (21.5%), *Capillaria* spp. (10.5%); trematoda: *Alaria alata* (1.6%) and cestodes from the Taeniidae family (1.3%). Factors like age, size and coat length, as well as the way of living, attitude and diet were linked to a significantly higher ($p < 0.05$) prevalence of intestinal parasites. Based on the results of coprological diagnostics, this research indicates the importance of educating dog owners, conducting routine parasitological tests on their pets and regular deworming strategies.

Keywords: dogs; helminths; protozoa; zoonoses; risk factors

1. Introduction

Among social animals, dogs are considered the most popular pets worldwide. Over the last decade, the interaction between humans and dogs has significantly increased, leading to these animals being treated as equal members of the family [1]. Such interactions may pose a risk of transmitting zoonotic pathogens. Dogs can be infected with different intestinal parasites, including protozoa (*Giardia intestinalis*, *Cystoisospora* spp., *Sarcocystis* spp., *Neospora caninum*) and helminths (roundworms, hookworms, whipworms and tapeworms) [2–9]. Clinical symptoms depend primarily of the dog's health, the type and severity of the parasite infection and the presence of additional parasitic infections in other organ systems (e.g., cardiorespiratory, urinary). However, infections are often asymptomatic.

Various studies conducted worldwide report a high prevalence of different types of parasites in the category of owned dogs [2,7,10–12]. Accordingly, they may serve as reservoirs of zoonotic parasites and can contaminate soil with the infectious stages of the parasites, such as the eggs and larvae of helminths, as well as the cysts of protozoa [13–17]. Some dog parasites can also infect humans, causing disease. Infection can occur either directly (i.e., trophically) and/or indirectly through contaminated food and water in the environment [18,19]. The risk of infection depends on various factors, both biological and environmental, which vary based on the parasite's life cycle and human behavior. Studies indicate that many pet owners are unaware of how dog endoparasites are transmitted and the public health risks they pose [10,11,20].

Knowing the epidemiological situation of intestinal parasites in dogs and identifying the ways they spread are key elements for effectively monitoring this threat. Bearing in mind the importance of veterinary medicine in maintaining the One Health concept, and recognizing the role of dogs in the spread of parasitic zoonoses, the aim of this study was to (i) identify gastrointestinal (GI) parasites in owned dogs using coprological diagnostics and (ii) to assess the risk factors important for the occurrence, maintenance and spread of parasitic infections.

2. Material and Methods

2.1. Study Area

The survey was conducted from November 2022 to June 2023 in seven administrative districts in the Republic of Serbia: Belgrade, Podunavlje, Kolubara, Mačva, West Bačka, Toplica and Bor (Figure 1). Serbia is a landlocked country located in the Balkan peninsula and the Pannonian Plain. Serbia lies between latitudes 41° and 47° N, and longitudes 18° and 23° E. In the northern part of the country, the climate is more continental, with colder winters and warmer summers, while in the southern part, the climate tends to be more Mediterranean, with milder winters and hotter summers. The average annual rainfall ranges from around 600 to 1000 mm. The average elevation of Serbia is approximately 500 m above sea level.

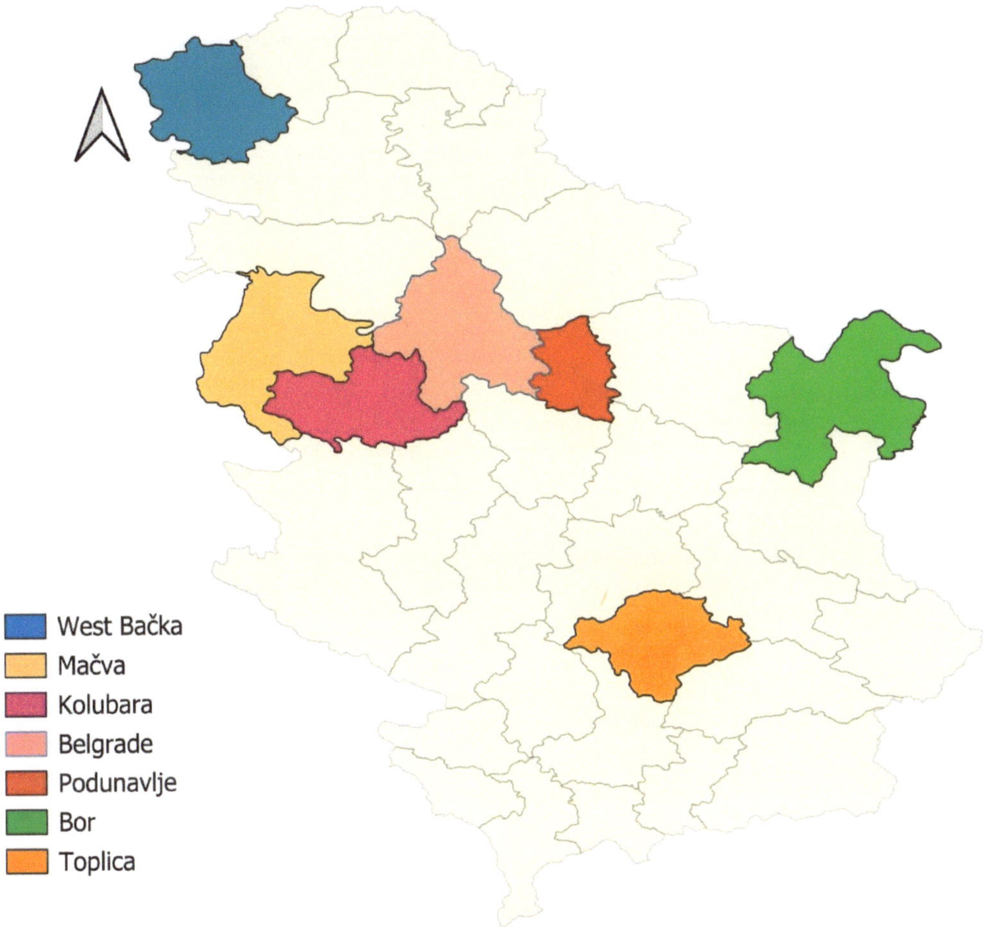

Figure 1. Map of Serbia with administrative districts where the survey was conducted. The map was generated by using QGIS v3.36 [21].

2.2. *Coproparasitological Examination*

A total of 382 fecal samples were collected from owned dogs. The samples were stored at +4 °C in labeled disposable containers and transported to the Department of Parasitology at the Faculty of Veterinary Medicine, University of Belgrade, for parasitological analysis. Coproparasitological examination included the assessment of samples using both macroscopic and microscopic methods. Macroscopic examination was used to evaluate the presence of adult nematodes and proglottids of tapeworms (described in Ilić et al. [3]). For microscopic examination, samples (approximately 5–10 g per sample) were prepared using qualitative coprological diagnostic procedures—centrifugal flotation with saturated zinc sulphate solution (with a specific density of 1.18 at 20 °C). Each fecal sample was examined in duplicate under a light microscope (Olympus CX 23, Olympus, Tokyo, Japan) at magnifications of 100× and 400×. All eggs found were photographed and identified according to their morphological characteristics [22].

2.3. *Risk Factors Assessment*

In this study, we investigated the influence of various individual factors and environmental variables. The analysis of individual variables encompassed the following parameters: sex (male or female), age (<1 year, 1–5 years, 5–10 years, >10 years), size (<25 kg and >25 kg) and coat length (short, medium and long hair). The analysis of environmental variables considered the following parameters: attitude (pet, hunting, guard), diet (commercial, mixed, combined), habitat (indoor, outdoor, indoor/outdoor) and contact with other animals (yes or no).

The category of "pet dogs" includes animals kept in households for companionship, as well as those under the owner's care with restricted movement. "Hunting dogs" are animals owned and maintained by hunters, assisting in locating, chasing, and recovering prey during hunting activities. The "guard dogs" category comprises dogs protecting property in yards, with controlled or partially controlled movement [12].

Commercial diets included branded foods designed to meet the nutritional requirements of pets for each stage of life or lifestyle. A mixed diet implied the consumption of different foods (such as raw meat, offal and bread) and access to paratenic or intermediate hosts. A combined diet included both, commercial and mixed diet.

2.4. *Statistical Analyses*

Results were analyzed using Graph Pad Prism software, version 7 (GraphPad, San Diego, CA, USA). Factors (individual and environmental variables) associated with parasitism were analyzed using the Chi-Square (X^2) test. The odds ratio (OR) was calculated to verify the level of risk associated with variables that correlated with parasitism. To calculate the odds ratio, the following formula was used: $p \pm Z\ (p \times (1 - p)/n) \times 0.5$, where p is prevalence, Z is the multiplier from the normal distribution at a 95% confidence interval (1.96) and n is the number of examined samples. In all analyses, the confidence level was 95%, and statistical analyses were considered significant if $p < 0.05$, $p < 0.01$ and $p < 0.001$.

3. Results

3.1. *Prevalence of Gastrointestinal Parasites*

Through copromicroscopic investigation of fecal samples, endoparasites were found with a total prevalence of 62.6% (239/382). The prevalence of infections caused by protozoa was 12.3% (47/382), helminths 37.7% (144/382) and co-infection with protozoa and helminths was 12.6% (48/382). Eleven different species, genera or families of intestinal parasites were detected (Figure 2). The most prevalent protozoa was *Giardia intestinalis* (11.8%, 45/382). The presence of oocysts of *Cystoisospora* spp. (9.2%, 35/382), *Sarcocystis* spp. (4.5%, 17/382) and *Neospora caninum*/*Hammondia* spp. (3.7%, 14/382) were also detected. Of the nematodes, the most prevalent were Ancylostomatidae (38.0%, 145/382) and *Trichuris vulpis* (21.5%, 82/382), followed by *Toxocara canis* (11.5%, 44/382), *Capillaria* spp. (10.5%, 40/382) and *Toxascaris leonina* (4.2%, 16/382). Low prevalence of the trematode *Alaria alata* (1.6%, 6/382) and cestodes from the family Taeniidae (1.3%, 5/382) were also diagnosed. The most prevalent were monoinfections of dogs (29.8%, 114/382), followed by infections with two (18.1%, 69/382), three (9.2%, 35/382), four (2.1%, 8/382), five (2.62%, 10/382) and six (0.3%, 1/382) types of parasites (Table 1).

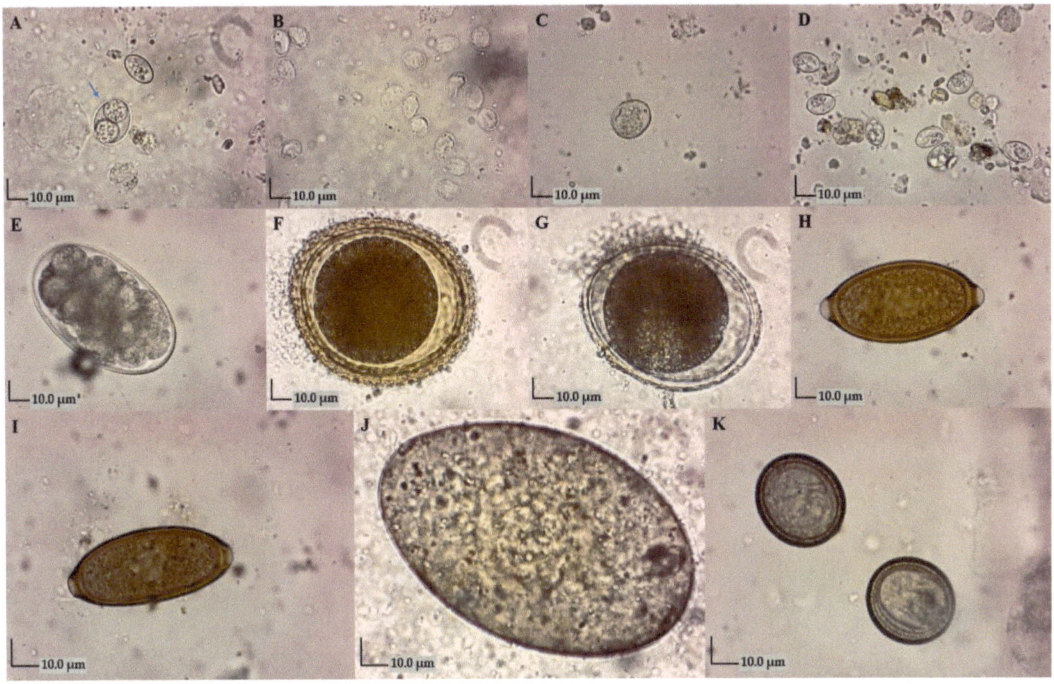

Figure 2. Parasitic elements detected in fecal samples, zinc sulphate flotation (×400): (**A**)—*Cystoisospora* spp. oocyst (blue arrow); (**B**)—*Giardia intestinalis* cysts; (**C**)—*Neospora caninum*/*Hammondia* spp. oocyst; (**D**)—*Sarcocystis* spp. sporocysts; (**E**)—Ancylostomatidae egg; (**F**)—*Toxocara canis* egg; (**G**)—*Toxascaris leonina* egg; (**H**)—*Trichuris vulpis* egg; (**I**)—*Capillaria* spp. egg; (**J**)—*Alaria alata* egg; (**K**)—Taeniidae eggs.

Table 1. Prevalence of intestinal parasites.

Endoparasites	n = 382		
	Positive Samples	%	**95% CI**
Cystoisospora spp.	35	9.2	6.30–12.10
Sarcocystis spp.	17	4.5	2.42–6.58
Neospora caninum/*Hammondia* spp.	14	3.7	1.81–5.69
Giardia intestinalis	45	11.8	8.56–15.04
Toxocara canis	44	11.5	8.30–14.70
Toxascaris leonina	16	4.2	2.10–6.20
Ancylostomatidae	145	38.0	33.13–42.87
Trichuris vulpis	82	21.5	17.38–25.62
Capillaria spp.	40	10.5	7.43–13.57
Alaria alata	6	1.6	0.34–2.86
Taeniidae	5	1.3	0.16–2.44
Occurrence of Infections	n = 382		
	Positive Samples	%	**95% CI**
Protozoa	47	12.3	9.01–15.59
Helminths	144	37.7	32.84–42.56
Protozoa + Helminths	48	12.6	9.06–16.14

Table 1. *Cont.*

Occurrence of Mixed Infections	n = 382		
	Positive Samples	**%**	**95% CI**
With one parasite	114	29.84	25.25–34.43
With two parasites	69	18.06	14.20–21.92
With three parasites	35	9.16	6.27–12.05
With four parasites	8	2.09	0.66–3.52
With five parasites	10	2.62	1.02–4.22
With six parasites	1	0.26	0–0.77

n—number of examined samples, CI—Confidence interval.

3.2. *Individual Fisk Factors*

The prevalence of endoparasitic infections was higher in male dogs (64.2%, 122/190) than in female dogs (60.9%, 117/192). Regarding the age of the dogs, a significantly higher prevalence of endoparasites ($p < 0.001$) was recorded in dogs younger than 1 year (83.3%, 55/66) compared to dogs aged 1–5 years (64.4%, 239/216), aged 5–10 years (46.2%, 36/78) and those older than 10 years (40.9%, 9/22) (Table 2). The prevalence of *G. intestinalis* (36.36%, 24/66), *T. canis* (27.27%, 18/64) and *T. leonina* (10.61%, 7/66) was significantly higher ($p < 0.05$; $p < 0.001$) in dogs <1 year, while a significantly higher ($p <$ 0.05) prevalence of Ancylostomatidae was found in dogs <1 year and 1–5 years (Table 3). Endoparasitic infections were significantly higher ($p < 0.05$) in dogs weighing less than 25 kg (67.2%, 154/229) compared to those weighing over 25 kg (55.6%, 85/153) (Table 2). A significantly higher ($p < 0.05$) prevalence of *T. canis* (14.41%, 33/229), *T. leonina* (6.11%, 14/229) and Ancylostomatidae (41.92%, 96/229) was recorded in dogs weighing less than 25 kg (Table 4). Prevalence of endoparasites was significantly higher ($p < 0.001$) in short-haired dogs (67.2%, 160/238) compared to medium-haired (60.7%, 54/89) and long-haired dogs (45.5%, 25/55) (Table 2). The prevalence of Ancylostomatidae (42.86%, 102/238) and *Capillaria* spp. (13.87%, 33/238) was significantly higher ($p < 0.05$) in short-haired dogs (Table 4).

Table 2. Individual and environmental risk factors.

		n	N	%	χ^2	p	**Odds Ratio**
Individual risk factors				Sex			
	Male	190	122	64.2	0.44	0.51	1.15
	Female	192	117	60.9			
				Size			
	<25 kg	229	154	67.2	5.36	*	1.64
	>25 kg	153	85	55.6			
				Age			
	<1 year	66	55	83.3	25.85	***	
	1–5 year	216	139	64.4			2.77
	5–10 year	78	36	46.2			5.83
	>10 year	22	9	40.9			7.22
				Coat length			
	Short	238	160	67.2	9.22	***	
	Medium	89	54	60.7			1.33
	Long	55	25	45.5			2.96

Table 2. *Cont.*

		n	N	%	χ^2	p	Odds Ratio
Environmental risk factors	Living with other animals						
	Yes	272	172	63.2	0.45	0.50	0.85
	No	110	67	60.9			
	Habitat						
	Indoor	55	9	16.4	59.17	***	
	Outdoor	127	86	67.7			0.09
	Indoor/Outdoor	200	144	72.0			0.07
	Diet						
	Commercial	86	40	46.5	19.53	***	
	Mixed food	200	145	72.5			0.33
	Combined	96	54	56.3			0.68
	Attitude						
	Pet	143	64	44.8	42.49	***	
	Guard dog	92	55	59.8			0.54
	Hunting dog	147	120	81.6			0.18

n—number of examined samples; *N*—number of positive samples; * $p < 0.05$; *** $p < 0.001$.

3.3. Environmental Risk Factors

Gastrointestinal parasites were more prevalent among dogs that were living with other animals (63.2%, 172/272) compared to those that were not (60.9%, 67/110) (Table 2). A significantly higher prevalence ($p < 0.05$, $p < 0.01$, $p < 0.001$) of *Cystoisospora* spp. (11.76%, 32/272), *Sarcocystis* spp. (5.88%, 16/272), *G. intestinalis* (15.07%, 41/272) and *T. canis* (14.34%, 39/272) was found among dogs that were living with other animals. On the contrary, a significantly higher prevalence ($p < 0.05$, $p < 0.01$) of Ancylostomatidae (50.0%, 55/110) and *Capillaria* spp. (15.45%, 17/110) was observed among dogs without contact with other animals (Table 5). Considering attitude, the prevalence of endoparasites was significantly higher ($p < 0.001$) among hunting dogs (81.6%, 120/147) compared to guard dogs (59.8%, 55/92) and pets (44.8%, 64/143) (Table 2). Among hunting dogs, a significantly higher prevalence ($p < 0.05$, $p < 0.01$, $p < 0.001$) was found for *Sarcocystis* spp. (9.52%, 14/147), *T. canis* (17.01%, 25/147), *T. leonina* (8.16%, 12/147), Ancylostomatidae (61.22%, 90/147), *T. vulpis* (34.01%, 50/147) and *Capillaria* spp. (17.69%, 26/147) (Table 5).

A significantly higher ($p < 0.001$) prevalence of parasites was recorded in the category of indoor/outdoor dogs (72.0%, 144/200) compared to outdoor (67.7%, 86/127) and indoor (16.4%, 9/55) dogs (Table 2). A significantly higher prevalence ($p < 0.05$; $p < 0.01$; $p < 0.001$) of *Cystoisospora* spp. (15.75%, 20/127), *Sarcocystis* spp. (8.66%, 11/127), *N. caninum/Hammondia* spp. (7.09%, 9/127) and *Alaria alata* (3.94%, 5/127) was found in the outdoor dog category. On the contrary, a significantly higher prevalence ($p < 0.05$; $p < 0.01$; $p < 0.001$) of *G. intestinalis* (17.5%, 35/200), Ancylostomatidae (45.5%, 91/200), *T. vulpis* (26.0%, 52/200) and *Capillaria* spp. (13.5%, 27/200) was found in the category of indoor/outdoor dogs (Table 6). In the category of dogs consuming mixed food (72.5%, 145/200), the prevalence of endoparasites was significantly higher ($p < 0.001$) compared to dogs consuming combined food (56.3%, 54/96) or commercial food (46.5%, 40/86) (Table 2). The prevalence of *Cystoisospora* spp. (13.5%, 27/200), *Sarcocystis* spp. (7.0%, 14/200), Ancylostomatidae (51.0%, 102/200) and *T. vulpis* (29.0%, 58/200) was significantly higher ($p < 0.05$; $p < 0.01$; $p < 0.001$) in dogs fed with a mixed diet, while the prevalence of *G. intestinalis* was significantly higher ($p < 0.05$) in dogs fed with a combined diet (Table 6).

Table 3. Influence of individual risk factors (sex and age) on prevalence of intestinal parasites.

n	Sex						Age									
	Male 229		Female 153		χ^2	*p*	<1 Year 66		1–5 Year 216		5–10 Year 78		>10 Year 22		χ^2	*p*
End	*N*	% (95% CI)	*N*	% (95% CI)			*N*	% (95% CI)	*N*	% (95% CI)	*N*	% (95% CI)	*N*	% (95% CI)		
Cys	19	10.00 (5.73–14.27)	16	8.33 (4.42–12.24)	0.32	0.57	11	16.67 (10.99–22.35)	19	8.80 (5.02–12.58)	4	5.13 (0.23–9.13)	1	4.55 (0–13.26)	6.59	0.09
Sar	8	4.21 (1.36–7.06)	9	4.69 (1.70–7.68)	0.05	0.82	2	3.03 (0–7.17)	10	4.63 (1.83–7.43)	5	6.41 (1.06–11.76)	0	0.00	2.06	0.56
Neo	7	3.68 (1–6.36)	7	3.65 (1–6.31)	0	1	2	3.03 (0–7.17)	7	3.24 (0.88–5.60)	3	3.85 (0–8.05)	2	9.09 (0–21.10)	2.03	2.03
Gia	22	11.58 (7.03–16.13)	23	11.98 (7.39–16.57)	0.02	0.91	24	36.36 (24.75–47.97)	18	8.33 (5.11–12.02)	2	2.56 (0–6.07)	1	4.55 (0–13.26)	48.33	***
Tox	21	11.05 (6.59–15.51)	23	11.98 (7.39–16.57)	0.08	0.78	18	27.27 (16.53–38.01)	21	9.72 (5.77–13.67)	4	5.13 (0.23–10.03)	1	4.55 (0–13.26)	20.93	***
Tas	5	2.63 (0.35–4.49)	11	5.73 (2.44–9.02)	2.28	0.13	7	10.61 (3.18–18.04)	7	3.24 (0.88–5.60)	2	2.56 (0–6.07)	0	0.00	8.73	*
Anc	75	39.47 (32.52–46.42)	70	36.46 (29.65–43.27)	0.37	0.54	27	40.91 (29.05–52.77)	93	43.06 (36.46–49.66)	20	25.64 (15.95–35.33)	5	22.73 (5.22–40.24)	9.82	*
Tri	41	21.58 (15.73–27.43)	41	21.35 (8.81–33.89)	0	1	13	19.70 (10.10–29.30)	49	22.69 (17.10–28.28)	16	20.51 (11.55–29.47)	4	18.18 (2.07–34.29)	0.50	0.92
Cap	22	11.58 (7.03–16.13)	18	9.38 (5.26–13.50)	0.50	0.48	8	12.12 (4.25–19.99)	28	12.96 (8.48–17.44)	4	5.13 (0.23–10.03)	0	0.00	6.57	0.09
Ala	3	1.58 (0–3.35)	3	1.56 (0–3.31)	0	1	1	1.52 (0–4.47)	2	0.93 (0–2.21)	2	2.56 (0–6.07)	1	4.55 (0–13.26)	2.34	0.51
Tae	3	1.58 (0–3.35)	2	1.04 (0–2.48)	0.21	0.64	0	0.00	5	2.31 (0.31–4.31)	0	0.00	0	0.00	3.89	0.27

n—number of examined samples; *N*—number of positive samples; CI—Confidence interval; * $p < 0.05$; *** $p < 0.001$; Cys—*Cystoisospora* spp.; Sar—*Sarcocystis* spp.; Neo—*Neospora caninum*/*Hammondia* spp.; Gia—*Giardia intestinalis*; Tox—*Toxocara canis*; Tas—*Toxascaris leonina*; Anc—Ancylostomatidae; Tri—*Trichuris vulpis*; Cap—*Capillaria* spp.; Ala—*Alaria alata*; Tae—Taeniidae.

Table 4. Influence of individual risk factors (size and coat length) on prevalence of intestinal parasites.

n	Size						Coat Length							
	<25 kg 229		>25 kg 153		χ^2	*p*	Short 238		Medium 89		Long 55		χ^2	*p*
End	*N*	% (95% CI)	*N*	% (95% CI)			*N*	% (95% CI)	*N*	% (95% CI)	*N*	% (95% CI)		
Cys	23	10.04 (6.15–13.93)	12	7.84 (3.54–12.14)	0.53	0.46	17	7.14 (3.77–10.51)	13	14.61 (7.27–21.94)	5	9.09 (1.49–16.69)	4.34	0.11
Sar	11	4.80 (2.03–7.57)	6	3.92 (0.84–7.00)	0.17	0.68	12	5.04 (2.26–7.82)	5	5.62 (0.83–10.40)	0	0.00	3.04	0.22
Neo	9	3.93 (1.41–6.45)	5	3.27 (0.45–6.09)	0.11	0.74	9	3.78 (1.36–6.20)	3	3.37 (0–7.12)	2	3.64 (0–8.59)	0.03	0.98
Gia	30	13.10 (8.70–17.50)	15	9.80 (5.09–14.51)	0.96	0.33	30	12.61 (8.39–16.83)	11	12.36 (5.52–19.22)	4	7.27 (0.41–14.13)	1.26	0.53

Table 4. *Cont.*

n	Size				χ^2	*p*	Coat Length						χ^2	*p*
	<25 kg 229		>25 kg 153				Short 238		Medium 89		Long 55			
End	*N*	% (95% CI)	*N*	% (95% CI)			*N*	% (95% CI)	*N*	% (95% CI)	*N*	% (95% CI)		
Tox	33	14.41 (9.86–18.96)	11	7.19 (3.10–11.28)	4.69	*	29	12.18 (8.02–16.34)	11	12.36 (5.52–19.22)	4	7.27 (0.41–14.13)	1.14	0.57
Tas	14	6.11 (3.01–9.21)	2	1.31 (0–3.11)	5.28	*	9	3.78 (1.36–6.20)	6	6.74 (1.53–11.95)	1	1.82 (0–5.35)	2.31	0.31
Anc	96	41.92 (35.53–48.31)	49	32.03 (24.64–39.42)	3.81	*	102	42.86 (36.57–49.15)	28	31.46 (21.81–41.61)	15	27.27 (15.50–39.04)	6.69	*
Tri	53	23.14 (17.68–28.60)	29	18.95 (12.74–15.16)	0.96	0.33	55	23.11 (17.75–28.47)	19	21.35 (12.84–29.86)	8	14.55 (5.23–23.87)	1.95	0.38
Cap	28	12.23 (7.99–16.47)	12	7.84 (3.54–12.14)	1.88	0.17	33	13.87 (9.48–18.26)	5	5.62 (0.83–10.40)	2	3.64 (0–8.59)	7.90	*
Ala	3	1.31 (0–2.78)	3	1.96 (0–4.16)	0	1	5	2.10 (0.27–3.91)	1	1.12 (0–3.31)	0	0.00	1.43	0.49
Tae	3	1.31 (0–2.78)	2	1.31 (0–3.11)	0	1	4	1.68 (0.05–3.31)	0	0.00	1	1.82 (0–5.35)	1.55	0.46

n—number of examined samples; *N*—number of positive samples; CI—Confidence interval; * $p < 0.05$; Cys—*Cystoisospora* spp.; Sar—*Sarcocystis* spp.; Neo—*Neospora caninum*/*Hammondia* spp.; Gia—*Giardia intestinalis*; Tox—*Toxocara canis*; Tas—*Toxascaris leonina*; Anc—Ancylostomatidae; Tri—*Trichuris vulpis*; Cap—*Capillaria* spp.; Ala—*Alaria alata*; Tae—Taeniidae.

Table 5. Influence of environmental risk factors (living with other animals and attitude) on prevalence of intestinal parasites.

n	Living with Other Animals				χ^2	*p*	Attitude						χ^2	*p*
	Yes 272		No 110				Pet 143		Guard Dog 92		Hunting Dog 147			
End	*N*	% (95% CI)	*N*	% (95% CI)			*N*	% (95% CI)	*N*	% (95% CI)	*N*	% (95% CI)		
Cys	32	11.76 (7.93–15.59)	3	2.73 (0–5.77)	7.69	**	12	8.39 (3.85–12.93)	7	7.61 (2.19–13.03)	16	10.88 (5.85–15.91)	12	0.64
Sar	16	5.88 (3.08–8.68)	1	0.91 (0–2.68)	4.56	*	2	1.40 (0–3.33)	1	1.09 (0–2.96)	14	9.52 (4.78–14.26)	2	***
Neo	11	4.04 (1.70–6.38)	3	2.73 (0–5.77)	0.39	0.53	6	4.20 (0.91–7.49)	4	4.35 (0.18–8.52)	4	2.72 (0.10–5.34)	6	0.74
Gia	41	15.07 (10.82–19.32)	4	3.64 (0.14–7.14)	9.86	***	18	12.59 (7.15–18.03)	13	14.13 (7.01–21.25)	14	9.52 (4.78–14.26)	18	0.52
Tox	39	14.34 (10.17–18.51)	5	4.55 (0.66–8.44)	7.37	**	10	6.99 (2.81–11.17)	9	9.78 (3.71–15.85)	25	17.01 (10.94–23.08)	10	*
Tas	12	4.41 (1.97–6.85)	4	3.64 (0.14–7.14)	0.12	0.73	1	0.70 (0–2.56)	3	3.26 (0–6.89)	12	8.16 (3.73–12.58)	1	**
Anc	90	33.09 (27.50–39.68)	55	50.00 (40.66–59.34)	9.51	**	23	16.08 (10.06–22.10)	32	34.78 (25.05–44.51)	90	61.22 (53.34–69.10)	23	***
Tri	54	19.85 (15.11–24.59)	28	25.45 (17.31–33.59)	1.46	0.23	14	9.79 (4.92–14.66)	18	19.57 (11.46–27.68)	50	34.01 (26.35–41.67)	14	***
Cap	23	8.46 (5.15–11.77)	17	15.45 (8.70–22.20)	4.09	*	7	4.90 (1.36–8.44)	7	7.61 (2.19–13.03)	26	17.69 (11.52–23.86)	7	***

Table 5. *Cont.*

n	Living with Other Animals				χ^2	*p*	Attitude						χ^2	*p*
	Yes 272		No 110				Pet 143		Guard Dog 92		Hunting Dog 147			
End	*N*	% (95% CI)	*N*	% (95% CI)			*N*	% (95% CI)	*N*	% (95% CI)	*N*	% (95% CI)		
Ala	6	2.21 (0.46–3.96)	0	0.00	2.47	0.12	1	0.70 (0–2.56)	2	2.17 (0–5.15)	3	2.04 (0–4.33)	1	0.57
Tae	3	1.10 (0–2.34)	2	1.82 (0–4.32)	0.31	0.58	1	0.70 (0–2.56)	0	0.00	4	2.72 (0.10–5.34)	1	0.14

n—number of examined samples; *N*—number of positive samples; CI—Confidence interval; * $p < 0.05$; ** $p < 0.01$; *** $p < 0.001$; Cys—*Cystoisospora* spp.; Sar—*Sarcocystis* spp.; Neo—*Neospora caninum*/*Hammondia* spp.; Gia—*Giardia intestinalis*; Tox—*Toxocara canis*; Tas—*Toxascaris leonina*; Anc—Ancylostomatidae; Tri—*Trichuris vulpis*; Cap—*Capillaria* spp.; Ala—*Alaria alata*; Tae—Taeniidae.

Table 6. Influence of environmental risk factors (habitat and diet) on prevalence of intestinal parasites.

n	Habitat						χ^2	*p*	Diet						χ^2	*p*
	Indoor 55		Outdoor 127		Indoor/Outdoor 200				Commercial 86		Mixed Food 382		Combined 96			
End	*N*	% (95% CI)	*N*	% (95% CI)	*N*	% (95% CI)			*N*	% (95% CI)	*N*	% (95% CI)	*N*	% (95% CI)		
Cys	1	1.82 (0–5.33)	20	15.75 (9.41–22.08)	14	7.00 (3.46–10.54)	11.31	**	3	3.49 (0–7.37)	27	13.5 (8.76–18.24)	5	5.21 (0.76–9.66)	9.65	***
Sar	0	0.00	11	8.66 (3.77–13.55)	6	3.00 (0.64–5.36)	8.85	*	0	0.00	14	7.0 (3.46–10.54)	3	3.13 (0–6.61)	7.46	*
Neo	0	0.00	9	7.09 (2.62–11.55)	5	2.50 (0.34–4.66)	7.07	*	3	3.49 (0–7.37)	11	5.5 (2.34–8.60)	0	0.00	5.57	0.06
Gia	4	7.27 (0.41–14.63)	6	4.72 (1.03–8.41)	35	17.50 (12.23–22.77)	13.46	***	16	18.60 (10.38–16.82)	15	7.5 (3.85–11.15)	14	14.58 (7.52–21.64)	8.11	*
Tox	3	5.45 (0–11.45)	17	13.39 (7.47–19.31)	24	12.00 (7.50–16.50)	2.46	0.29	7	8.14 (2.37–14.92)	27	13.5 (8.76–18.24)	10	10.42 (5.54–15.30)	1.85	0.40
Tas	1	1.82 (0–5.33)	5	3.94 (0.56–7.32)	10	5.00 (1.98–8.02)	1.12	0.57	4	4.65 (0.20–9.10)	9	4.5 (1.63–6.57)	3	3.13 (0–6.61)	0.37	0.83
Anc	2	3.64 (0–8.59)	52	40.94 (32.39–49.49)	91	45.50 (38.55–52.40)	32.82	***	16	18.60 (10.38–16.82)	102	51.0 (44.07–57.93)	27	28.13 (19.14–37.13)	32.07	***
Tri	1	1.82 (0–5.33)	29	22.83 (15.53–30.13)	52	26.00 (19.92–32.08)	15.18	***	5	5.81 (0.87–10.75)	58	29.0 (22.71–35.23)	19	19.79 (11.82–27.76)	19.39	***
Cap	0	0.00	13	10.24 (4.97–15.51)	27	13.50 (8.76–18.24)	8.40	*	8	9.30 (3.16–15.44)	25	12.5 (7.92–17.08)	7	7.29 (2.09–12.49)	2.04	0.36
Ala	0	0.00	5	3.94 (0.56–7.32)	1	0.50 (0–1.48)	6.96	*	1	1.16 (0–3.42)	3	1.5 (0–3.18)	2	2.08 (0–4.93)	0.26	0.88
Tae	0	0.00	2	1.57 (0–3.73)	3	1.50 (0–3.18)	0.86	0.65	0	0.00	4	2.0 (0.06–3.94)	1	1.04 (0–3.07)	1.93	0.38

n—number of examined samples; *N*—number of positive samples; CI—Confidence interval; * $p < 0.05$; ** $p < 0.01$; *** $p < 0.001$; Cys—*Cystoisospora* spp.; Sar—*Sarcocystis* spp.; Neo—*Neospora caninum*/*Hammondia* spp.; Gia—*Giardia intestinalis*; Tox—*Toxocara canis*; Tas—*Toxascaris leonina*; Anc—Ancylostomatidae; Tri—*Trichuris vulpis*; Cap—*Capillaria* spp.; Ala—*Alaria alata*; Tae—Taeniidae.

4. Discussion

In our research, the total prevalence of gastrointestinal parasites in owned dogs was 62.6%. This finding is in accordance with previous research on dogs in public shelters in Serbia [3], which reported a total GI parasite prevalence of 58.3%. The results of numerous studies conducted in European countries reveal the different prevalence of endoparasites in dogs. In Greece [23,24], the prevalence ranged from 26% to 65%, in Slovakia from 27.1% to 45.7% [12,25], in Spain 53.6% [26], in Portugal from 41.0 to 81.19% [27–29] and in Germany 41.2% [30]. From the total number of examined fecal samples, the most frequent findings were monoinfections (29.8%), followed by infections with two (18.1%), three (9.2%), four (2.1%), five (2.62%), and six (0.3%) endoparasites. Similar to our findings, other studies have reported monoinfections as the most prevalent, while polyparasitism was also confirmed [8,12,27,31–34]. The prevalence of infections caused by protozoa in dogs in this research was 12.3%, helminths 37.7%, and co-infections with both protozoa and helminths 12.6%. A study from Spain found a higher prevalence of helminths (63.6%) in hunting dogs compared to intestinal protozoa (20.4%). In contrast, dogs from shelters had a higher prevalence of intestinal protozoa (67.9%) than helminths (9.8%) [35]. The heterogeneity of the available results depends on the origin of the samples (farm dogs, hunting dogs, owned dogs, shelter dogs, stray dogs) and the socio-economic status of the countries where the research was carried out [20].

4.1. Protozoa

Among the protozoa, *Giardia intestinalis* was the most prevalent (11.8%). It is widely reported in both domestic and wild animals, which can serve as hosts and reservoirs of zoonotic Assemblages [36–39]. This parasite is among the most common in humans, with an estimated 200 million people infected [40]. The prevalence of giardiosis in humans in developed countries ranges between 2 and 7%, and in developing countries 20 and 30% [41]. In this research, *G. intestinalis* was the most prevalent protozoa in dogs younger than one year (36.36%). Our results are in accordance with the results in the study by Murnik et al. [30], where the prevalence of *G. intestinalis* was 29%. An increased risk of giardiosis in dogs younger than one year has been confirmed in studies by other authors [42–45]. A higher prevalence was detected among the category of guard dogs and pets, as well as those who lived indoors/outdoors. Additionally, dogs that were fed commercial or combined diets and were in contact with other animals had a higher prevalence. Given the various ways *G. intestinalis* can spread through contaminated food and water [46–48], it is clear that these specific groups of dogs can serve as a source of environmental contamination, posing an indirect threat to individuals, particularly farmers, veterinarians, and animal handlers [41].

Oocysts of *Cystoisospora* spp. were identified in 9.2% of the samples examined. Oocysts were found most frequently in dogs younger than one year. The higher prevalence of *Cystoisospora* spp. found in younger dogs was confirmed in our previous study [3]. These results are also in accordance with Papazahariadou et al. [23], who reported a significantly higher number of coccidiosis cases in young dogs compared to adults. In addition, a higher prevalence of *Cystoisospora* spp. was found in dogs that live outside, have contact with other animals and consume mixed diets. This finding may be associated with the contaminated environment and the presence of this protozoa in the soil [15].

Among protozoa, a lower prevalence of *Sarcocystis* spp. (4.5%) and *Neospora caninum/Hammondia* spp. (3.7%) was found. The highest prevalence of *Sarcocystis* spp. was diagnosed in the category of hunting dogs (9.52%), which is not in accordance with results from Germany, where a high prevalence of sarcocystosis (63.3%) was found in hunting dogs in areas inhabited by wolves [49]. In that research, prevalence was determined using molecular methods, which is a more sensitive method than microscopical examination. Such differences could be explained by the assumption that the investigated hunting dogs originated from areas where wolves live. Compared to pet dogs in Germany, where the prevalence of sarcocystosis ranged from 2 to 9% [50], we found a lower prevalence in both pet dogs (1.40%) and guard dogs (1.09%) in our study. A higher prevalence was found in dogs that live outdoors, have contact with other animals and consume mixed diets. Such dogs are allowed to feed on the meat of herbivores, which are intermediate hosts for these protozoa, thus maintaining the circulation of this parasite [49,51].

Oocysts of *N. caninum/Hammondia* spp. were the most frequent in the category of dogs living outside (7.09%). Dogs fed a mixed diet had the highest number of positive samples (5.5%). This is likely because these dogs have the opportunity to consume infected tissues (raw or undercooked meat, fetal membranes) or intermediate hosts containing tissue cysts [52]. Given that *N. caninum* can cause abortions in cattle and cause economic loses in livestock, this category of dogs may pose a risk to cattle health. This risk is supported by the findings of Klun et al. [53], who reported a seroprevalence of this coccidia of 7.2% in cattle in Serbia.

4.2. *Nematoda*

The most common GI parasites identified in our study were hookworms from the Ancylostomatidae family (38.0%). This finding is in accordance with results from Bulgaria, where these parasites were most prevalent in owned and stray dogs, dogs that live outside and harbor dogs [54–56]. In our previous study on dogs from public shelters [3], the prevalence of Ancylostomatidae was 15.4%. A significantly higher prevalence of parasites was found in the category of dogs younger than one year and aged from 1 to 5 years, short-haired dogs and dogs lighter than 25 kg. Also, a higher prevalence of these nematodes was found in hunting dogs, dogs fed mixed diets and those living indoors/outdoors. Our finding aligns with Letra Mateus et al. [27], who reported a high prevalence of Ancylostomatidae in hunting dogs. This could be due to the dogs being kept together in groups and creating a favorable environment for parasite transmission. Also, factors such as hunting prey and consuming a wider variety of food sources might contribute to a higher risk of infection [9]. Rubel et al. [57] reported that the prevalence of hookworm is higher in regions with lower socio-economic status. On the contrary, in a study in Germany, in dogs younger than one year, the prevalence of these parasites was 0.9% [30]. Nematodes from the Ancylostomatidae family pose a risk to human health, given that their infectious stage can cause cutaneous larva migrans, and in the case of *Ancylostoma caninum*, eosinophilic enteritis or neuroretinitis [58–60].

Trichuris vulpis was the second most common parasite and was diagnosed in 21.5% of examined dogs. In research conducted in Bulgaria, this nematode was found in 15.1% of owned dogs kept outdoors [56] and 20% of dogs from shelters [54]. Additionally, it was found in 13.6% of dogs from shelters in Italy [32] and in 20% of dogs in Romania [61]. A lower prevalence was observed in 9.5% of owned dogs in Albania [62], 9.6% of hunting and herding dogs in Greece [23] and 4.8% of domestic dogs, along with 13.6% of shelter dogs in Italy [32]. In research conducted in Spain [35] and Portugal [27], trichuriosis was the most prevalent in hunting dogs, similar to our study (34.01%). A higher prevalence of *T. vulpis* was found in the category of dogs using mixed and combined food. The eggs of these parasites can remain viable for years, contaminating the environment, food and water, thereby posing a risk for infections in dogs [63].

Toxocara canis was found in 11.5% of the examined samples, with the highest prevalence in the population of hunting dogs (17.01%) and in dogs younger than one year (27.27%). The obtained results are consistent with findings from Europe, where *T. canis* prevalence ranged from 17.72% in Spain [64] to 11.9% to 16.5% in Slovakia [13,65], 12.8% in Greece [23], 5.1% to 11.28% in Portugal [27,28], 8% in Albania [62] and 6.4% in Bulgaria [56]. Comparing this with previous research conducted in Serbia, a higher prevalence of toxocarosis in pet dogs was observed at 16.6% [2], while in owned dogs that visit public parks it ranged from 36.6% to 38% [4], and in dogs from shelters it was 33.5% [3]. The larvae of this ascarid may pose a risk to humans, as upon infection they migrate into internal organs, potentially leading to visceral and ocular larva migrans [66]. In this regard, Deutz et al. [67] confirmed a high seroprevalence of *T. canis* among farmers, slaughterhouse staff, veterinarians and hunters.

Eggs of the trichurid type, exhibiting morphological characteristics specific to species from the genus *Capillaria*, were diagnosed in 10.5% of the fecal samples, with the assumption that they belong to a species of *C. aerophila*. The prevalence of *C. aerophila* in dogs across Europe and the Balkan countries varies, ranging from 0.4% to 0.5% in Italy [32], 0.65% in Romania [61], 2.8% in Albania [62] and from 2% to 11% in Bulgaria [55]. The prevalence of respiratory capillariosis in dogs cannot be determined with certainty, as the excreted eggs may not exclusively originate from adult parasites inhabiting the trachea. They could also appear in feces due to coprophagia or ingestion of food previously contaminated with eggs of *Capillaria* spp. from the feces of other dogs or animals [62]. A Higher prevalence of *Capillaria* spp. was found in the category of hunting dogs, those who live outside and those in contact with other animals. These results are not surprising, since a higher prevalence of *C. aerophila* (38%) was found in red foxes in Serbia [68].

The prevalence of *Toxascaris leonina* species in owned dogs was 4.2%. The highest number of positive findings was observed in hunting dogs (17.01%) and in the category of dogs younger than one year (10.61%). Ilić et al. [3] reported a prevalence of toxascarosis of 3.4% in dogs from public shelters in Serbia, while authors from Slovakia found this ascarid in 1.6% of various categories of dogs [12].

4.3. *Trematoda and Cestoda*

Among the other parasites, a lower prevalence of the trematode *Alaria alata* (1.6%) was diagnosed in this study. Besides wild carnivores, which are definitive hosts and contribute to the spread of *A. alata* [69], this parasite was confirmed in our study among hunting and guard dogs, as well as in dogs that live outdoors. The presence of *A. alata* was also found in the category of dogs that were fed with a mixed and combined diet. However, one positive sample was recorded in a dog fed commercial food, suggesting that the infection occurred after the consumption of intermediate hosts while the dog was outside.

Cestodes from the family Taeniidae were confirmed in five dogs (1.3%), which is slightly lower than the prevalence found in different categories of dogs (4%) in Slovakia [12] and in Germany (up to 12.2%) [30,49]. The positive samples were mostly obtained from hunting dogs that frequently stay in the wild during hunting, which is why they are at a higher risk of consuming intermediate hosts [27,70]. Although the eggs of species from the family Taeniidae cannot be differentiated by light microscopy, in veterinary medicine, as a precaution, any eggs of the taeniid type found are considered as the presence of eggs of the species *Echinococcus granulosus*. The presence of *E. granulosus* in the feces of owned dogs is particularly important for public health.

5. Conclusions

In the research, the total prevalence of endoparasites was 62.6%. Of particular importance for public health is the discovery of the largest number of gastrointestinal parasites found in categories of dogs younger than one year, hunting dogs, dogs kept indoors/outdoors and those fed with mixed food. Considering the finding of zoonotic endoparasites and the presence of species with zoonotic potential, the obtained results are particularly important for owners and veterinarians in clinical practice. These findings can aid in the adequate selection of antiparasitics, planning of deworming regimens and implementation of programs for the prevention of parasitic infections in dogs.

Author Contributions: Conceptualization, N.M.J. and T.I.; Data curation, K.N. and M.M. (Milorad Mirilovic); Investigation, N.M.J., O.B., D.B., M.R. and M.M. (Milan Maletic); Project administration, T.I.; Supervision, T.I.; Writing—original draft, N.M.J., O.B. and M.R.; Writing—review and editing, D.B. and T.I. All authors have read and agreed to the published version of the manuscript.

Funding: The study was supported by the Ministry of Science, Technological Development and Innovation of the Republic of Serbia (Contract number 451-03-66/2024-03/200143).

Institutional Review Board Statement: Approval (No. 01-03/2024) of this study was obtained from the Ethical Commission for the Protection of Animal Welfare, Faculty of Veterinary Medicine, University of Belgrade.

Informed Consent Statement: Not applicable.

Data Availability Statement: The data presented in this study are available on request from the corresponding author.

Acknowledgments: The authors extend their sincere appreciation to the doctors of veterinary medicine for their invaluable cooperation and support throughout this study.

Conflicts of Interest: The authors declare no conflicts of interest.

References

1. McConnell, A.R.; Brown, C.M.; Shoda, T.M.; Stayton, L.E.; Martin, C.E. Friends with benefits: On the positive consequences of pet ownership. *J. Personal. Soc. Psychol.* **2011**, *101*, 1239. [CrossRef]
2. Ilic, T.; Kulisic, Z.; Antić, N.; Radisavljevic, K.; Dimitrijevic, S. Prevalence of zoonotic intestinal helminths in pet dogs and cats in the Belgrade area. *J. Appl. Anim. Res.* **2017**, *45*, 204–208. [CrossRef]
3. Ilic, T.; Nisavic, U.; Gajic, B.; Nenadovic, K.; Ristic, M.; Stanojevic, D.; Dimitrijevic, S. Prevalence of intestinal parasites in dogs from public shelters in Serbia. *Comp. Immunol. Microbiol. Infect. Dis.* **2021**, *76*, 101653. [CrossRef] [PubMed]
4. Ristic, M.; Dimitrijevic, S.; Visnjic, A.; Bogunovic, D.; Gajic, B.; Stojanovic, M.; Ilic, T. Dogs from public city parks as a potential source of pollution of the environment and risk factor for human health. *Indian J. Anim. Res.* **2020**, *90*, 535–542. [CrossRef]
5. Ferreira, J.I.G.D.S.; Pena, H.F.J.; Azevedo, S.S.; Labruna, M.B.; Gennari, S.M. Occurrences of gastrointestinal parasites in fecal samples from domestic dogs in São Paulo, SP, Brazil. *Rev. Bras. Parasitol. Vet.* **2016**, *25*, 435–440. [CrossRef] [PubMed]
6. Sobotyk, C.; Upton, K.E.; Lejeune, M.; Nolan, T.J.; Marsh, A.E.; Herrin, B.H.; Borst, M.M.; Piccione, J.; Zajac, A.M.; Camp, L.E.; et al. Retrospective study of canine endoparasites diagnosed by fecal flotation methods analyzed across veterinary parasitology diagnostic laboratories, United States, 2018. *Parasites Vectors* **2021**, *14*, 439. [CrossRef] [PubMed]
7. Kamani, J.; Massetti, L.; Olubade, T.; Balami, J.A.; Samdi, K.M.; Traub, R.J.; Colella, V.; González-Miguel, J. Canine gastrointestinal parasites as a potential source of zoonotic infections in Nigeria: A nationwide survey. *Prev. Vet. Med.* **2021**, *192*, 105385. [CrossRef] [PubMed]
8. Safarov, A.; Mihalca, A.D.; Park, G.M.; Akramova, F.; Ionică, A.M.; Abdinabiev, O.; Deak, G.; Azimov, D. A survey of helminths of dogs in rural and urban areas of Uzbekistan and the zoonotic risk to human population. *Pathogens* **2022**, *11*, 1085. [CrossRef] [PubMed]
9. Mendoza Roldan, J.A.; Otranto, D. Zoonotic parasites associated with predation by dogs and cats. *Parasites Vectors* **2023**, *16*, 55. [CrossRef]
10. Figueredo, L.A.; da Silva Sales, K.G.; Deuster, K.; Pollmeier, M.; Otranto, D.; Dantas-Torres, F. Exposure to vector-borne pathogens in privately owned dogs living in different socioeconomic settings in Brazil. *Vet. Parasitol.* **2017**, *243*, 18–23. [CrossRef]
11. Lempereur, L.; Nijsse, R.; Losson, B.; Marechal, F.; De Volder, A.; Schoormans, A.; Martinelle, L.; Danlois, F.; Claerebout, E. Coprological survey of endoparasite infections in owned dogs and owners' perceptions of endoparasite control in Belgium and the Netherlands. *Vet. Parasitol. Reg. Stud. Rep.* **2020**, *22*, 100450. [CrossRef] [PubMed]
12. Jarošová, J.; Antolová, D.; Lukáč, B.; Madari, A. A Survey of intestinal helminths of dogs in Slovakia with an emphasis on zoonotic species. *Animals* **2021**, *11*, 3000. [CrossRef] [PubMed]
13. Papajová, I.; Pipiková, J.; Papaj, J.; Čižmár, A. Parasitic contamination of urban and rural environments in the Slovak Republic: Dog's excrements as a source. *Helminthologia* **2014**, *51*, 273–280. [CrossRef]
14. Simonato, G.; Cassini, R.; Morelli, S.; Di Cesare, A.; La Torre, F.; Marcer, F.; Traversa, D.; Pietrobelli, M.; di Regalbono, A.F. Contamination of Italian parks with canine helminth eggs and health risk perception of the public. *Prev. Vet. Med.* **2019**, *172*, 104788. [CrossRef] [PubMed]
15. Bogunovic, D.; Dominikovic, N.; Jovanovic, N.; Nenadovic, K.; Kulisic, Z.; Ilic, T.; Stevic, N. Environmental contamination by parasites in public parks in Belgrade in the context of one health approach. *Acta Vet. Belgrade* **2022**, *72*, 30–44. [CrossRef]
16. Ristic, M.; Miladinovic-Tasic, N.; Dimitrijevic, S.; Nenadovic, K.; Bogunovic, D.; Stepanovic, P.; Ilic, T. Soil and sand contamination with canine intestinal parasite eggs as a risk factor for human health in public parks in Niš (Serbia). *Helminthologia* **2020**, *57*, 109–119. [CrossRef]
17. Lorenzo-Rebenaque, L.; López-Fernández, S.; Marco-Jiménez, F.; Montoro-Dasi, L.; Marin, C.; Vega, S.; Martínez-Manzanares, E.; Fariñas, F. Zoonotic Parasites in Playgrounds in Southern Spain: A One Health Approach. *Microorganisms* **2023**, *11*, 721. [CrossRef] [PubMed]
18. Esch, K.J.; Petersen, C.A. Transmission and epidemiology of zoonotic protozoal diseases of companion animals. *Clin. Microbiol. Rev.* **2013**, *26*, 58–85. [CrossRef] [PubMed]
19. Majewska, A.A.; Huang, T.; Han, B.; Drake, J.M. Predictors of zoonotic potential in helminths. *Phil. Trans. R. Soc. B* **2021**, *376*, 20200356. [CrossRef]

20. Otranto, D.; Dantas-Torres, F.; Mihalca, A.D.; Traub, R.J.; Lappin, M.; Baneth, G. Zoonotic parasites of sheltered and stray dogs in the era of the global economic and political crisis. *Trends Parasitol.* **2017**, *33*, 813–825. [CrossRef]
21. QGIS Geographic Information System (Version 3.36). Open Source Geospatial Foundation Project. 2024. Available online: http://qgis.org (accessed on 23 February 2024).
22. Hendrix, C. *Diagnostic Parasitology for Veterinary Technicians*, 2nd ed.; Penn Foster: St. Louis, MO, USA, 1998.
23. Papazahariadou, M.; Founta, A.; Papadopoulos, E.; Chliounakis, S.; Antoniadou-Sotiriadou, K.; Theodorides, Y. Gastrointestinal parasites of shepherd and hunting dogs in the Serres Prefecture, Northern Greece. *Vet. Parasitol.* **2007**, *148*, 170–173. [CrossRef]
24. Diakou, A.; Di Cesare, A.; Morelli, S.; Colombo, M.; Halos, L.; Simonato, G.; Tamvakis, A.; Beugnet, F.; Paoletti, B.; Traversa, D. Endoparasites and vector-borne pathogens in dogs from Greek islands: Pathogen distribution and zoonotic implications. *PLoS Negl. Trop. Dis.* **2019**, *13*, e0007003. [CrossRef] [PubMed]
25. Szabová, E.; Juriš, P.; Miterpáková, M.; Antolová, D.; Papajová, I.; Šefčíková, H. Prevalence of important zoonotic parasites in dog populations from the Slovak Republic. *Helminthologia* **2007**, *44*, 170–176. [CrossRef]
26. Benito, A.; Carmena, D.; Postigo, I.; Estibalez, J.J.; Martinez, J.; Guisantes, J.A. Intestinal helminths in dogs in Alava, North of Spain. *Res. Rev. Parasitol.* **2003**, *63*, 21–126.
27. Letra Mateus, T.; Castro, A.; Niza Ribeiro, J.; Vieira-Pinto, M. Multiple zoonotic parasites identified in dog feces collected in Ponte de Lima, Portugal—A potential threat to human health. *Int. J. Environ. Res. Public Health* **2014**, *11*, 9050–9067. [CrossRef] [PubMed]
28. Neves, D.; Lobo, L.; Simões, P.B.; Cardoso, L. Frequency of intestinal parasites in pet dogs from an urban area (Greater Oporto, Northern Portugal). *Vet. Parasitol.* **2014**, *200*, 295–298. [CrossRef] [PubMed]
29. Cardoso, A.S.; Costa, I.M.H.; Figueiredo, C.; Castro, A.; Conceição, M.A.P. The occurrence of zoonotic parasites in rural dog populations from northern Portugal. *J. Helminthol.* **2014**, *88*, 203–209. [CrossRef]
30. Murnik, L.C.; Daugschies, A.; Delling, C. Gastrointestinal parasites in young dogs and risk factors associated with infection. *Parasitol. Res.* **2023**, *122*, 585–596. [CrossRef]
31. Xhaxhiu, D.; Kusi, I.; Rapti, D.; Kondi, E.; Postoli, R.; Rinaldi, L.; Dimitrova, Z.M.; Visser, M.; Knaus, M.; Rehbein, S. Principal intestinal parasites of dogs in Tirana, Albania. *Parasitol. Res.* **2011**, *108*, 341–353. [CrossRef]
32. Scaramozzino, P.; Carvelli, A.; Iacoponi, F.; De Liberato, C. Endoparasites in household and shelter dogs from Central Italy. *Int. J. Vet. Sci. Med.* **2018**, *6*, 45–47. [CrossRef]
33. Raza, A.; Rand, J.; Qamar, A.G.; Jabbar, A.; Kopp, S. Gastrointestinal parasites in shelter dogs: Occurrence, pathology, treatment and risk to shelter workers. *Animals* **2018**, *8*, 108. [CrossRef] [PubMed]
34. Souza, J.B.B.; Silva, Z.M.D.A.; Alves-Ribeiro, B.S.; Moraes, I.D.S.; Alves-Sobrinho, A.V.; Saturnino, K.C.; Ferraz, H.T.; Machado, M.R.F.; Braga, Í.A.; Ramos, D.G.D.S. Prevalence of intestinal parasites, risk factors and zoonotic aspects in dog and cat populations from Goiás, Brazil. *Vet. Sci.* **2023**, *10*, 492. [CrossRef] [PubMed]
35. Ortuño, A.; Scorza, V.; Castellà, J.; Lappin, M. Prevalence of intestinal parasites in shelter and hunting dogs in Catalonia, Northeastern Spain. *Vet. J.* **2014**, *199*, 465–467. [CrossRef] [PubMed]
36. Ballweber, L.R.; Xiao, L.; Bowman, D.D.; Kahn, G.; Cama, V.A. Giardiasis in dogs and cats: Update on epidemiology and public health significance. *Trends Parasitol.* **2010**, *26*, 180–189. [CrossRef]
37. Feng, Y.; Xiao, L. Zoonotic potential and molecular epidemiology of *Giardia* species and giardiasis. *Clin. Microbiol. Rev.* **2011**, *24*, 110–140. [CrossRef] [PubMed]
38. Ryan, U.; Cacciò, S.M. Zoonotic potential of Giardia. *Int. J. Parasitol.* **2013**, *43*, 943–956. [CrossRef]
39. Thompson, R.C.A.; Ash, A. Molecular epidemiology of *Giardia* and *Cryptosporidium* infections. *Infect. Genet. Evol.* **2016**, *40*, 315–323. [CrossRef]
40. Certad, G.; Viscogliosi, E.; Chabé, M.; Cacciò, S.M. Pathogenic mechanisms of *Cryptosporidium* and *Giardia*. *Trends Parasitol.* **2017**, *33*, 561–576. [CrossRef] [PubMed]
41. Dixon, B.R. *Giardia duodenalis* in humans and animals—Transmission and disease. *Res. Vet. Sci.* **2021**, *135*, 283–289. [CrossRef]
42. Palmer, C.S.; Traub, R.J.; Robertson, I.D.; Devlin, G.; Rees, R.; Thompson, R.A. Determining the zoonotic significance of *Giardia* and *Cryptosporidium* in Australian dogs and cats. *Vet. Parasitol.* **2008**, *154*, 142–147. [CrossRef]
43. Batchelor, D.J.; Tzannes, S.; Graham, P.A.; Wastling, J.M.; Pinchbeck, G.L.; German, A.J. Detection of endoparasites with zoonotic potential in dogs with gastrointestinal disease in the UK. *Transbound. Emerg. Dis.* **2008**, *55*, 99–104. [CrossRef] [PubMed]
44. Gates, M.C.; Nolan, T.J. Endoparasite prevalence and recurrence across different age groups of dogs and cats. *Vet. Parasitol.* **2009**, *166*, 153–158. [CrossRef] [PubMed]
45. Itoh, N.; Kanai, K.; Kimura, Y.; Chikazawa, S.; Hori, Y.; Hoshi, F. *Giardia* infections in pet shop puppies in 2008 and 2013. *J. Vet. Intern. Med.* **2015**, *24*, 32–34.
46. Budu-Amoako, E.; Greenwood, S.J.; Dixon, B.R.; Barkema, H.W.; McClure, J.T. Foodborne illness associated with *Cryptosporidium* and *Giardia* from livestock. *J. Food Prot.* **2011**, *74*, 1944–1955. [CrossRef] [PubMed]
47. Santin, M. *Cryptosporidium* and *Giardia* in ruminants. *Vet. Clin. N. Am. Food Anim. Pract.* **2020**, *36*, 223–238. [CrossRef]
48. Cirkovic, V.; Klun, I.; Utaaker, K.S.; Uzelac, A.; Tysnes, K.R.; Robertson, L.J.; Djurkovic-Djakovic, O. Surface waters as a potential source of *Giardia* and *Cryptosporidium* in Serbia. *Exp. Parasitol.* **2020**, *209*, 107824. [CrossRef] [PubMed]
49. Lesniak, I.; Franz, M.; Heckmann, I.; Greenwood, A.D.; Hofer, H.; Krone, O. Surrogate hosts: Hunting dogs and recolonizing grey wolves share their endoparasites. *Int. J. Parasitol. Parasites Wildl.* **2017**, *6*, 278–286. [CrossRef]

50. Barutzki, D.; Schaper, R. Results of parasitological examinations of faecal samples from cats and dogs in Germany between 2003 and 2010. *Parasitol. Res.* **2011**, *109*, 45–60. [CrossRef]
51. Thompson, R.A. Parasite zoonoses and wildlife: One health, spillover and human activity. *Int. J. Parasitol.* **2013**, *43*, 1079–1088. [CrossRef]
52. Silva, R.C.; Machado, G.P. Canine neosporosis: Perspectives on pathogenesis and management. *Vet. Med. Res.* **2016**, *26*, 59–70. [CrossRef]
53. Klun, I.; Cirkovic, V.; Maletic, M.; Bradonjic, S.; Djurkovic-Djakovic, O. Seroprevalence of *Neospora caninum* infection and associated risk factors in dairy cattle in Serbia. *Parasitol. Res.* **2019**, *118*, 1875–1883. [CrossRef] [PubMed]
54. Radev, V.; Lalkovski, N.; Zhelyazkov, P.; Kostova, T.; Sabev, P.; Nedelchev, N.; Vassileva, R. Prevalence of gastrointestinal parasites and *Dirofilaria* spp. in stray dogs from some regions in Bulgaria. *Bulg. J. Vet. Med.* **2016**, *19*, 57–62. [CrossRef]
55. Iliev, P.; Kirkova, Z.; Ivanov, A.; Prelezov, P.; Tonev, A.; Kalkanov, I. Retrospective analysis on helminthic and protozoan infections in dogs and cats in Bulgaria. *Bulg. J. Vet. Med.* **2017**, *20*, 389–393.
56. Iliev, P.T.; Kirkova, Z.T.; Tonev, A.S. Preliminary study on the prevalence of endoparasite infections and vector-borne diseases in outdoor dogs in Bulgaria. *Helminthologia* **2020**, *57*, 171–178. [CrossRef] [PubMed]
57. Rubel, D.; Zunino, G.; Santillán, G.; Wisnivesky, C. Epidemiology of *Toxocara canis* in the dog population from two areas of different socioeconomic status, Greater Buenos Aires, Argentina. *Vet. Parasitol.* **2003**, *115*, 275–286. [CrossRef] [PubMed]
58. Hochedez, P.; Caumes, E. Hookworm-related cutaneous larva migrans. *J. Travel Med.* **2007**, *14*, 326–333. [CrossRef] [PubMed]
59. Rodriguez-Morales, A.J.; González-Leal, N.; Montes-Montoya, M.C.; Fernández-Espíndola, L.; Bonilla-Aldana, D.K.; Azeñas-Burgoa, J.M.; de Medina, J.C.D.; Rotela-Fisch, V.; Bermudez-Calderon, M.; Arteaga-Livias, K.; et al. Cutaneous larva migrans. *Curr. Trop. Med. Rep.* **2021**, *8*, 190–203. [CrossRef] [PubMed]
60. Hawdon, J.M.; Wise, K.A. Ancylostoma caninum and Other Canine Hookworms. In *Dog Parasites Endangering Human Health Parasitology Research Monographs*; Springer: Cham, Switzerland, 2021; pp. 147–193. [CrossRef]
61. Ursache, A.L.; Györke, A.; Mircean, V.; Matei, I.A.; Cozma, V. Prevalence of *Toxocara canis* and other endoparasites in client-owned dogs from Cluj County, Romania. *Sci. Parasitol.* **2017**, *17*, 101–107.
62. Shukullari, E.; Rapti, D.; Visser, M.; Pfister, K.; Rehbein, S. Parasites and vector-borne diseases in client-owned dogs in Albania: Infestation with arthropod ectoparasites. *Parasitol. Res.* **2017**, *116*, 399–407. [CrossRef]
63. Traversa, D. Are we paying too much attention to cardio-pulmonary nematodes and neglecting old-fashioned worms like *Trichuris vulpis*? *Parasites Vectors* **2011**, *4*, 32. [CrossRef]
64. Martínez-Moreno, F.J.; Hernández, S.; López-Cobos, E.; Becerra, C.; Acosta, I.; Martínez-Moreno, A. Estimation of canine intestinal parasites in Cordoba (Spain) and their risk to public health. *Vet. Parasitol.* **2007**, *143*, 7–13. [CrossRef] [PubMed]
65. Ondriska, F.; Mačuhová, K.; Melicherová, J.; Reiterová, K.; Valentová, D.; Beladičová, V.; Halgoš, J. Toxocariasis in urban environment of western Slovakia. *Helminthologia* **2013**, *50*, 261–268. [CrossRef]
66. Overgaauw, P.A.; van Knapen, F. Veterinary and public health aspects of *Toxocara* spp. *Vet. Parasitol.* **2013**, *193*, 398–403. [CrossRef]
67. Deutz, A.; Fuchs, K.; Auer, H.; Kerbl, U.; Aspöck, H.; Köfer, J. *Toxocara*-infestations in Austria: A study on the risk of infection of farmers, slaughterhouse staff, hunters and veterinarians. *Parasitol. Res.* **2005**, *97*, 390–394. [CrossRef] [PubMed]
68. Lalošević, V.; Lalošević, D.; Čapo, I.; Simin, V.; Galfi, A.; Traversa, D. High infection rate of zoonotic *Eucoleus aerophilus* infection in foxes from Serbia. *Parasite* **2013**, *20*, 3. [CrossRef] [PubMed]
69. Möhl, K.; Große, K.; Hamedy, A.; Wüste, T.; Kabelitz, P.; Lücker, E. Biology of *Alaria* spp. and human exposition risk to *Alaria* mesocercariae—A review. *Parasitol. Res.* **2009**, *105*, 1–15. [CrossRef]
70. Waindok, P.; Raue, K.; Grilo, M.L.; Siebert, U.; Strube, C. Predators in northern Germany are reservoirs for parasites of One Health concern. *Parasitol. Res.* **2021**, *120*, 4229–4239. [CrossRef]

animals

Article

Occurrence and Risk Factors Associated with *Platynosomum illiciens* Infection in Cats with Elevated Liver Enzymes

Pinkarn Chantawong [1,2], Jiraporn Potiwong [1], Natchanon Choochote [1], Kakanang Piyarungsri [1,3], Chakorn Kunkaew [4], Sahatchai Tangtrongsup [1,3,5,*,†] and Saruda Tiwananthagorn [1,2,3,*,†]

1 Faculty of Veterinary Medicine, Chiang Mai University, Chiang Mai 50100, Thailand; pinkarn_@hotmail.com (P.C.); potiwong.pim@gmail.com (J.P.); nonchoochote@gmail.com (N.C.); kakananjp@gmail.com (K.P.)
2 Research Center for Veterinary Biosciences and Veterinary Public Health, Faculty of Veterinary Medicine, Chiang Mai University, Chiang Mai 50100, Thailand
3 Research Center of Producing and Development of Products and Innovations for Animal Health and Production, Faculty of Veterinary Medicine, Chiang Mai University, Chiang Mai 50100, Thailand
4 Department of Animal Science and Fisheries, Faculty of Science and Agricultural Technology, Rajamangala University of Technology Lanna, Lampang 52000, Thailand; vetbird9@yahoo.com
5 Elephant, Wildlife, and Companion Animals Research Group, Faculty of Veterinary Medicine, Chiang Mai University, Chiang Mai 50100, Thailand
* Correspondence: sahatchai.t@cmu.ac.th (S.T.); saruda.t@cmu.ac.th (S.T.); Tel.: +66-53-948-015 (S.T.); +66-95-446-5955 (S.T.)
† These authors contributed equally to this work.

Simple Summary: Platynosomiasis is a feline hepatobiliary disease caused by *Platynosomum illiciens*, the most significant and widely distributed trematode in tropical and subtropical areas. In this study, we investigated the occurrence of *Platynosomum* spp. infection in cats with elevated serum alanine aminotransferase (ALT) levels and confirmed the species through DNA sequencing. Additionally, we assessed the association of factors and clinicopathological abnormalities with *Platynosomum* spp. infection to raise awareness and emphasize the importance of an appropriate deworming regimen to reduce the risk of *P. illiciens* infection.

Abstract: *Platynosomum* spp., a hepatic trematode, causes fatal hepatobiliary disease in cats. Feline platynosomiasis is often underestimated due to a lack of awareness and diagnostic challenges. This study aimed to investigate the occurrence, factors, and clinicopathological abnormalities associated with *Platynosomum* spp. infection in cats with elevated serum ALT levels. *Platynosomum* infection was determined using zinc sulfate flotation and formalin–ether sedimentation. DNA sequence analysis of PCR products from the *Platynosomum* internal transcribed spacer 2 (ITS2) region and *cox1* gene was used to identify *Platynosomum* species. Of a total of 43 cat fecal samples, the proportion of *Platynosomum* spp. infection by microscopic examination was 11.63% (5/43). All PCR-positive samples were molecularly identified as *Platynosomum illiciens*. From the logistic regression analysis, the odds of *Platynosomum* infection in cats without a deworming program were 16 times higher than those of regularly dewormed cats. Demographic data, housing conditions, and predatory behavior were not significantly associated with the infection. Regarding blood profiles, infected cats had higher eosinophil counts ($p = 0.014$), with no significant differences in ALT ($p = 0.791$) or ALP ($p = 0.970$) levels compared to non-infected cats. Our findings demonstrate that eosinophilia in cats with increased serum ALT may suggest *P. illiciens* infection in endemic areas. We strongly recommend a regular deworming program to mitigate the risk of *P. illiciens* infection.

Keywords: *Platynosomum illiciens*; liver fluke; cats; liver enzymes; associated factors

1. Introduction

Over the decades, helminthic infections in cats have gained the scientific interest of researchers due to the severity of their clinical manifestations [1]. The hepatic trematode *Platynosomum illiciens* (syn. *P. fastosum* and *P. concinnum*) is the most important and widely distributed parasite of cats in tropical and subtropical areas. The prevalence of this liver fluke has been reported to range from 0.07% to 81% worldwide [2]. In Asian countries, liver fluke infection in cats has been reported in Malaysia [3], Sri Lanka [4], Vietnam [5,6], Thailand [7–9], and Korea [10].

Feline platynosomiasis is considered a fatal disease; however, the information on this parasitic infection is scattered. Cats become infected by ingesting intermediate hosts (terrestrial isopods) or paratenic hosts (lizards, amphibians, and insects) containing metacercaria, which then migrate through the common bile duct, smaller biliary ducts, and the gall bladder, causing hepatobiliary problems such as cholangiohepatitis, cholangiocarcinoma, hepatic fibrosis, and bile duct obstruction, resulting in related clinicopathologic abnormalities [2,11]. Eosinophilia and an increased activity of the hepatic enzymes alanine aminotransferase (ALT) and alkaline phosphatase (ALP) have been detected in platynosomiasis cases. However, increased serum ALP concurrent with eosinophilia is sporadically reported [11–13]. Infected cats present a variety of clinical signs such as cachexia, vomiting, diarrhea, ascites, progressive jaundice, and death. Nonetheless, some cats remain asymptomatic. These signs vary according to the severity of the infection, the number of adult parasites, the duration of infection, and the immune response [14,15].

Platynosomum spp. infection can be diagnosed by coproparasitological evaluation, including zinc sulfate flotation and formalin–ether sedimentation techniques [16,17]. Additionally, molecular techniques such as polymerase chain reaction (PCR) assays and DNA sequence analyses of multiple target genes have been applied to identify *Platynosomum* species [18]. However, most cases of platynosomiasis have been diagnosed as incidental findings during necropsies due to its non-specific clinical effects [2,11]. This problem may lead to the underdiagnosis of this parasite and a lack of prompt treatment. To our knowledge, many reports have documented sudden deaths in platynosomiasis cases presenting only clinical pathology abnormalities, with a notable increase in hepatic enzymes [2,13,19]. Therefore, this study aimed to investigate the occurrence, risk factors, and clinical pathology abnormalities associated with *Platynosomum* infection in cats with elevated serum ALT levels. This fundamental study could increase awareness of this fatal disease among veterinary practitioners.

2. Materials and Methods

2.1. Ethical Approval

The research protocols were approved by the animal ethics committee of the Faculty of Veterinary Medicine, Chiang Mai University (9 July 2021; Ref. No. S19/2564), under the guidelines for the Care and Use of Experimental Animals, National Research Council of Thailand.

2.2. Study Area and Animal Selection

The study area was Chiang Mai Province in the northern region of Thailand, at the coordinates 18°47′47.22″ N, 98°57′40.644″ E (Figure 1), and samples were collected at the Small Animal Veterinary Teaching Hospital, Faculty of Veterinary Medicine. Domestic cats with elevated serum ALT levels (over 97 U/L, reference range 25–97 U/L) [20] and without a history of acute traumatic conditions were selected from this hospital between January 2020 and October 2021. Inclusion criteria were as follows: cats aged over three months, regardless of breed, sex, pregnant or lactating status, and cat's housing. No cats were treated with topical or systemic parasiticides in the 8 weeks preceding the study. The sample size of 43 cat fecal samples was calculated based on a 2.7% prevalence rate from a previous study conducted in an animal refuge in Nakhon Nayok, Thailand [8], with an acceptable relative error of 5% and a 95% confidence interval. The samples were randomly selected from cat populations to increase detection potential using a list of random numbers generated from the Open Epi program 3.01 [21].

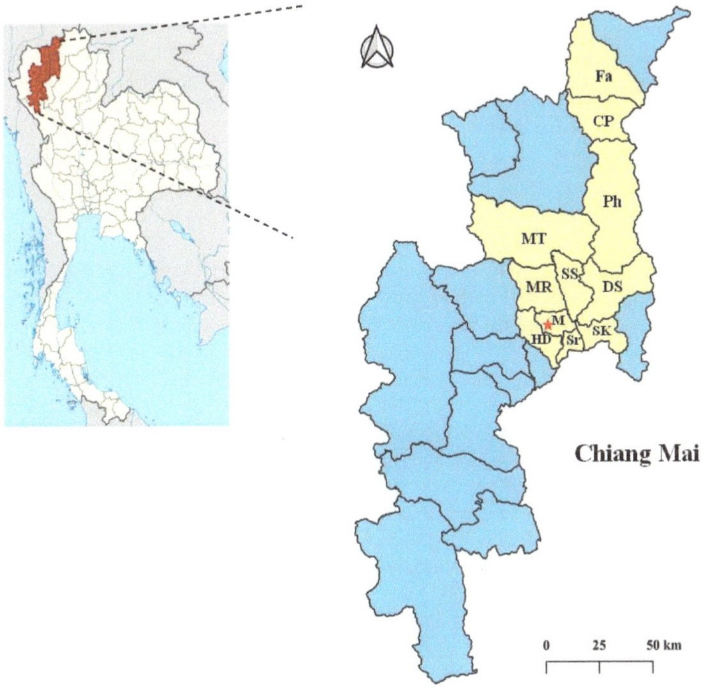

Figure 1. Map of the study area in Chiang Mai Province, Thailand. The star indicates the Small Animal Veterinary Teaching Hospital, Faculty of Veterinary Medicine. The 43 selected cats that reached the inclusion criteria resided in 11 districts, including Fang (Fa), Chai Prakan (CP), Phrao (Ph), Mae Tang (MT), Mae Rim (MR), San Sai (SS), Doi Saket (DS), Mueang Chiang Mai (M), Hang Dong (HD), Saraphi (Sr), and San Kamphaeng (SK).

2.3. Data and Sample Collection

The individual demographic data (age, sex, and breed) of 43 selected cats, as well as risk factors including the cats' housing (indoor or outdoor), predatory behavior, and deworming regimens (time and frequency of deworming), were retrieved from the cats' owners using a questionnaire. Clinical signs and blood profiles for each selected cat were recorded, including ALT, ALP, and eosinophil levels. Eosinophil count was performed using an automated hematological analyzer (BC-5300 Vet, Mindray, Shenzhen, China), and serum ALT and ALP assessments were performed using an automated chemistry analyzer (BX-3010, Sysmex Corporation, Tokyo, Japan) according to the manufacturer's recommendations.

Forty-three fecal samples were collected from the litter boxes of individually housed cats. After that, fecal samples were transported on ice to the Parasitology Laboratory at the Faculty of Veterinary Medicine, Chiang Mai University, Chiang Mai, Thailand. The samples were stored at 4 °C and analyzed within 1 week.

2.4. Parasitological Procedures and Identification

2.4.1. Fecal Analysis

All fecal samples were examined under a light microscope after simple flotation using zinc sulfate solution (specific gravity: 1.35) and formalin–ether sedimentation techniques [1,17]. A fecal sample was considered *Platynosomum*-positive if at least one *Platynosomum*-like egg, primarily identified using Basu et al.'s morphological keys [2], was identified. The remaining positive fecal samples were stored at −20 °C until further molecular analysis.

2.4.2. Molecular Technique for Species Identification

Genomic DNA samples from *Platynosomum*-positive fecal samples were extracted using the NucleoSpin® DNA stool kit (Macherey-Nagel GmbH, Düren, Germany) according to the manufacturer's instructions.

Two PCR assays were conducted to identify the species of *Platynosomum* in the positive samples. DNA fragments of the internal transcribed spacer 2 (ITS2) region (~560 bp) and trematode mitochondrial cytochrome c oxidase 1 (*cox1*) gene (396 bp) were amplified using published primers and thermocycler programs [18,22]. A forward primer, ITS5 (5′-GAAGTAAAAGTCGTAACAAGG-3′), and a reverse primer, d58R (5′-CACGAGCCGAGTGATCCACCGC-3′), were used for the ITS2 amplification, and a forward primer, JB3 (5′-TTTTTTGGGCATCCTGAGGTTTAT-3′), and a reverse primer, Trem.*cox1*.rrnl (5′-AATCATGATGCAAAAGGTA-3′), for *cox1* amplification. An annealing temperature of 55 °C was used to amplify the ITS2 fragment [18], and 50 °C was used for the *cox1* fragment [22]. Distilled water (DW) and adult *Platynosomum illiciens* DNA harvested from an infected cat carcass were used as negative and positive controls, respectively. PCR-positive products were purified using a Nucleospin® PCR Clean-up Kit (Macherey-Nagel GmbH, Düren, Germany) and submitted for direct sequencing using a commercially available direct sequencing service (Macrogen, Seoul, Republic of Korea) in the forward and reverse directions using the PCR primer sets described above. Only two *cox1* and four ITS2 nucleotide sequences of *Platynosomum* were successfully retrieved. Nucleotide sequences from forward and reverse directions were edited and manually aligned to retrieve a consensus sequence using BioEdit Sequence Alignment Editor Software version 7.7 [23]. Phylogenetic analysis was performed using MEGA X [24]. Available sequences of *Platynosomum* spp. from Vietnam [6], Brazil [18], and Sri Lanka [4] were used as reference sequences for tree construction. In addition, a sequence of *Dipylidium caninum* from Italy (MT806359) [25] and *Dicrocoelium dendriticum* from Japan (LC629058) [26] were used as an outgroup species for the *cox1* gene and ITS2 region, respectively. Multiple sequences were aligned using ClustalW, and phylogenetic analysis was performed using a maximum likelihood (ML) method based on the Tamura–Nei model [27]. A consensus tree was obtained after a 1000-replication bootstrap analysis.

2.5. Statistical Analysis

Demographic data, factors, clinical signs, and blood profiles were entered into a spreadsheet, and all statistical analyses were performed using Stata statistical software release 16.1 (StataCorp, College Station, TX, USA). The prevalence of *Platynosomum* infection in cats with elevated ALT levels was calculated with a 95% confidence interval (CI). Odds ratios (OR) and the 95% CI were estimated to measure the strength of the association using univariable logistic regression analysis. Variables associated with *Platynosomum* infection ($p \leq 0.1$) were included in a multivariable logistic regression analysis. A multivariable logistic regression model was performed using a backward stepwise elimination procedure against the parasitic examination results. The normality of serums ALT and ALP and eosinophil levels was assessed using the Shapiro–Wilk test. Differences in ALT, ALP, and eosinophil levels between parasitized and non-parasitized cats were compared using the Mann–Whitney U test. p-values ≤ 0.05 were considered statistically significant.

3. Results

3.1. Occurrence of Platynosomum spp. Infection

A total of 43 fecal samples were obtained from cats of different ages, sexes, and breeds (Table 1). The study population comprised 65.12% female and 34.88% male cats; most of them were juvenile (1–7 years old, 72.1%) and domestic short-haired cats (74.42%). The proportion of *Platynosomum* spp. infection in cats with elevated ALT levels was 11.63% (95% CI: 4.70–25.58), and all were determined as microscopically positive in both the zinc sulfate flotation and formalin–ether sedimentation techniques. The single infection rate of this hepatic trematode was 4.65%, whereas dual infections with *Toxocara* spp. or *Ancylostoma* spp. occurred at 2.33%, and triple infections with *Toxocara* spp. and *Toxascaris* spp. were found at 2.33% (Table 2).

Table 1. The demographic data (age, sex, and breed) of selected cat samples.

Demographic Data	Number (%)
Age	
Kitten (<1 year)	7 (16.28)
Juvenile (1–7 years)	31 (72.10)
Adult (7–14 years)	4 (9.30)
Geriatric (>14 years)	1 (2.33)
Sex	
Female	28 (65.12)
Neutered	19 (67.86)
Unneutered	9 (32.14)
Male	15 (34.88)
Neutered	6 (40.00)
Unneutered	9 (60.00)
Breed	
Domestic short-haired	32 (74.42)
Siamese	1 (2.32)
Persian	7 (16.28)
Scottish fold	2 (4.65)
British shorthair	1 (2.32)

Table 2. Occurrence of *Platynosomum* spp. infection and co-infection of other helminths assessed by microscopic examination.

Helminthic Infections	No. Positive (% Positive)
Overall helminthic infections	14 (32.56)
Platynosomum spp.	5 (11.63)
Ancylostoma spp.	3 (6.98)
Toxocara spp.	8 (18.60)
Toxascaris spp.	2 (4.65)
Spirometra spp.	1 (2.33)
Single helminthic infection	10 (23.26)
Platynosomum spp.	2 (4.65)
Ancylostoma spp.	2 (4.65)
Toxocara spp.	5 (11.63)
Spirometra spp.	1 (2.33)
Dual helminthic infections	3 (6.98)
Platynosomum spp. + *Toxocara* spp.	1 (2.33)
Platynosomum spp. + *Ancylostoma* spp.	1 (2.33)
Toxocara spp. + *Toxascaris* spp.	1 (2.33)
Triple helminthic infections	1 (2.33)
Platynosomum spp. + *Toxocara* spp. + *Toxascaris* spp.	1 (2.33)

The *Platynosomum* eggs presented with a mean length of 32.34 ± 10.85 μm (range: 13.75–40.0 μm) and a width of 22.6 ± 7.27 μm (range: 10.0–30.0 μm). A varied size of typical eggs was predominantly observed in the fecal samples of four infected cats. Morphologically, they were golden to dark brown, elliptical, and embryonated with an operculum at one end (Figure 2A–D). In one infected cat, an atypical egg was smaller and lacked an embryo (Figure 2E).

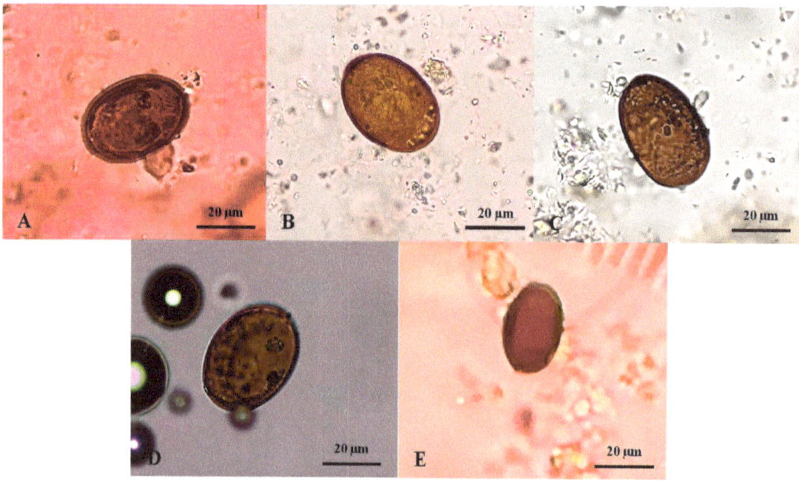

Figure 2. *Platynosomum* spp. eggs from infected cat fecal samples assessed by simple flotation method and examined under a light microscope. (**A–D**) *Platynosomum* spp. eggs were operculated, brown, and elliptical, containing undefined and granular embryos, and possibly germ balls inside. (**E**) Another form of *Platynosomum* spp. egg was smaller in size and brown, with a thick shell, long elliptical, and no internal germ balls.

3.2. *Platynosomum Species Confirmation and Genetic Characteristics of P. illiciens Based on ITS2 Region and cox1 Gene*

Four of the five *Platynosomum*-egg-positive samples were assessed using PCR assays targeting the ITS2 region and the *cox1* gene. One positive fecal sample was not available for PCR assays (Cat13-CMU). *Platynosomum* DNA was successfully amplified in all samples using ITS2 PCR assays, and in two samples using a *cox1* PCR assay. Nucleotide sequence analyses revealed that all *Platynosomum*-positive samples were *P. illiciens*. The sequences of *P. illiciens* in cat isolates were deposited in GenBank (DDBJ/EMBL/GenBank database Accession No. LC779845–LC779848 for ITS2 and LC779849–LC779850 for *cox1*). The *cox1* sequence in one cat (Cat1-CMU) exhibited 100% identity with *P. illiciens* found in a cat isolate from Brazil (GenBank: OM368257). The ITS2 sequence presented 99.74% identity with *P. illiciens* isolates from cats from Sri Lanka (GenBank: OK254044) and Costa Rica (GenBank: LC500530). As displayed in Figure 3A, the *cox1* sequence of one isolate, P1-COX1-Cat1-CMU, was placed in the same group of *P. illiciens* isolates from cats from Brazil (GenBank: MH155181-184), while P16-COX1-Cat16-CMU was separated in the different clade. Regarding the ITS2 sequences (Figure 3B), P9-ITS2-Cat9-CMU was close to *P. illiciens* isolates from cats from Brazil (GenBank: MH156565). Two isolates, P2-ITS2-Cat2-CMU and P16-ITS2-Cat16-CMU, were close to *P. illiciens* isolates from cats from Sri Lanka (GenBank: OK254044). One isolate, P1-ITS2-Cat1-CMU, was separated into a different clade.

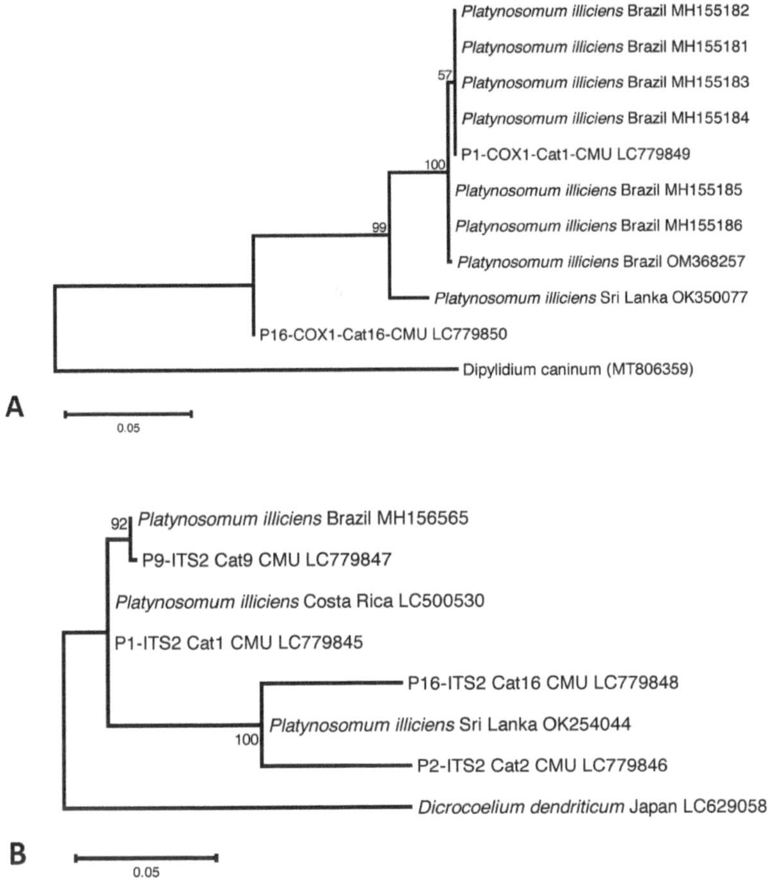

Figure 3. The evolutionary relationship of *P. illiciens* isolates from cats, based on 396-bp fragments of the mitochondrial cytochrome c oxidase subunit1 (*cox1*) gene (**A**), and on 396-bp fragments of the internal transcribed spacer 2 (ITS2) region (**B**). The tree was constructed using a maximum likelihood method based on the Tamura–Nei model using MEGA X software 10.2.6. The number in each branch indicates the percentage of 1000 bootstrap replications. Sequences obtained from GenBank are indicated by their accession numbers.

3.3. Associations of Demographic Data, Clinical Signs, and Factors with Platynosomum Infection

The associations of demographic data, clinical signs, factors, and *Platynosomum* infection are summarized in Table 3. Univariable logistic regression analyses demonstrated no significant association of *Platynosomum* infection with age, sex, breed, cat's housing, predatory behavior, or clinical signs, including anorexia, weakness, vomiting, and jaundice. On the other hand, the deworming program ($p = 0.003$) and deworming time ($p = 0.003$) were statistically significantly associated with the parasitic infection. For the multivariable logistic regression analysis, when controlling for jaundice, the odds of Platynosomum infection in cats without a deworming program were 16 times higher than those of regularly dewormed cats (Table 4). Interestingly, two out of five infected cats had a history of receiving deworming drugs but followed an irregular deworming program.

Table 3. Univariable analysis of the associations between factors and *Platynosomums* infection in cats with increased alanine aminotransferase (ALT) (n = 43).

Variable	*Platynosomum*-Positive Number (%)	Odds Ratio	95% CI	p Value
Sex				0.643
Male	1/15 (6.70)	Reference		
Female	4/28 (14.3)	2.33	0.24–23.00	0.468
Neuter status				0.717
Female—intact	1/9 (11.10)	Reference		
Female—spayed	3/19 (15.80)	1.48	0.10–88.45	≅1.000
Male—intact	0/9 (0.00)	1.00	0–39.00	≅1.000
Male—castrated	1/6 (16.70)	1.55	0.02–141.06	≅1.000
Age				0.771
Kitten (<1 year)	0/7 (0.00)	Reference		
Juvenile (1–7 years)	5/31 (16.1)	1.67	0.20-inf.	0.677
Adult (7–14 years)	0/4 (0.00)	1.00	0.00-inf.	
Geriatric (>14 years)	0/1 (0.00)	1.00	0.00-inf.	
Breed				0.738
Persian	0/7 (0.00)	Reference		
Domestic short-haired	5/32 (15.60)	1.61	0.19-inf.	0.700
Siamese	0/1 (0.00)	1.00	0.00-inf.	
Scottish fold	0/1 (0.00)	1.00	0.00-inf.	
British shorthair	0/2 (0.00)	1.00	0.00-inf.	
Cat's housing				0.589
Indoor	3/32 (9.40)	Reference		
Outdoor	2/11 (18.20)	2.11	0.15–21.62	0.760
Predatory behavior				0.145
Yes	5/28 (17.9)	4.09	0.15-inf.	0.204
No	0/15 (0.00)	Reference		
Deworming program				0.003
Regular	0/28 (0.00)	Reference		
Irregular	2/8 (25.00)	9.43	0.70-inf.	0.089
No deworming	3/7 (42.90)	20.68	2.00-inf.	0.011
Deworming time				0.003
2–6 months	0/24 (0.00)	Reference		
6–12 months	1/10 (10.00)	2.40	0.06-inf.	0.588
>12 months	1/2 (50.00)	12.00	0.31-inf.	0.154
No deworming	3/7 (42.90)	17.70	1.71-inf.	0.016
Clinical signs				≅1.000
No	3/25 (12.00)	Reference		
Yes	2/18 (11.11)	0.92	0.07–9.02	≅1.000
Anorexia				0.575
No	3/33 (9.09)	Reference		
Yes	2/10 (20.00)	2.44	0.18–25.41	0.657
Vomiting				≅1.000
No	5/37 (11.63)	Reference		
Yes	0/6 (0.00)	0.86	0–7.49	0.906
Jaundice				0.060
No	3/39 (7.69)	Reference		
Yes	2/4 (50.00)	10.72	0.59–202.65	0.120
Weakness				0.116
No	4/42 (9.52)	Reference		
Yes	1/1 (100.00)	7.60	0.19-Inf	0.233

Table 4. Multivariable analysis of the associations between factors and Platynosomums infection in cats with increased alanine aminotransferase (ALT) (n = 43).

Factor	Odds Ratio	95%CI	p Value
Irregular deworming	5.10	0.30-Inf	0.259
No deworming	15.95	1.40-Inf	0.022
Jaundice	6.13	0.23–522.33	0.439

Regarding the systemic deworming drugs, praziquantel at 5 mg/kg combined with pyrantel embonate at 57.5 mg/kg was the most commonly used parasiticide (34.88%; 15/43), followed by pyrantel embonate at 5 mg/kg (9.3%; 4/43) and fenbendazole at 30 mg/kg (4.65%; 2/43). Spot-on parasiticides were sporadically used, including fipronil at 11.11–33.2 mg/kg combined with (S)-methoprene at 13.51–40 mg/kg and praziquantel at 11.11–33.2 mg/kg (9.3%; 4/43); and imidacloprid at 12.5–25 mg/kg combined with moxidectin 1.25–2.5 mg/kg (4.65%; 2/43). Additionally, nine of the forty-three cats (20.93%) were dewormed with unknown medications.

3.4. Comparisons of Blood Profiles between Cats with and without Platynosomum Infection

Comparisons of serums ALT and ALP and eosinophil levels in parasitized and non-parasitized cats are shown in Figure 4. In platynosomiasis cats, the median of ALT was 134 U/L (IQR: 101–773 U/L), ALP was 80 U/L (IQR: 50–80 U/L), and eosinophil was 1.45×10^3 cells/μL (IQR: 1.00–3.27 $\times 10^3$ cells/μL); in non-infected cats, the median of ALT was 199 U/L (IQR: 115–277 U/L), ALP was 75.5 U/L (IQR: 52–101 U/L), and eosinophil was 0.57×10^3 cells/μL (IQR: 0.37–1.04 $\times 10^3$ cells/μL). Notably, the median eosinophil level in infected cats was significantly higher than in non-infected cats (p = 0.014). However, the two groups had no significant differences in levels of ALT (p = 0.791) and ALP (p = 0.970).

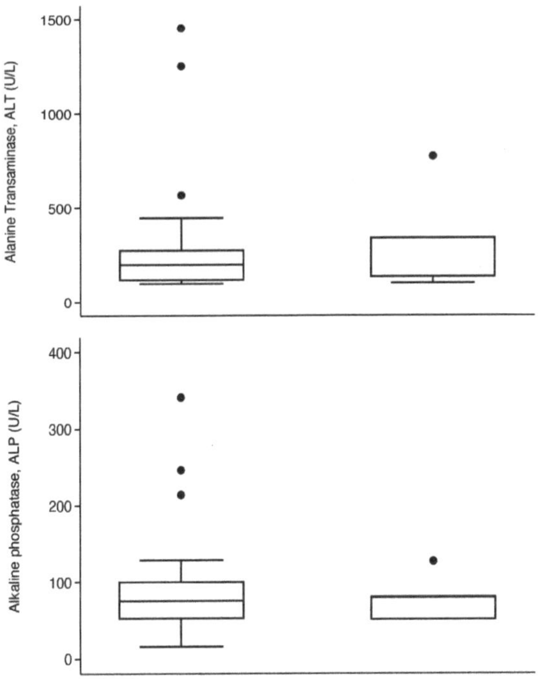

Figure 4. *Cont.*

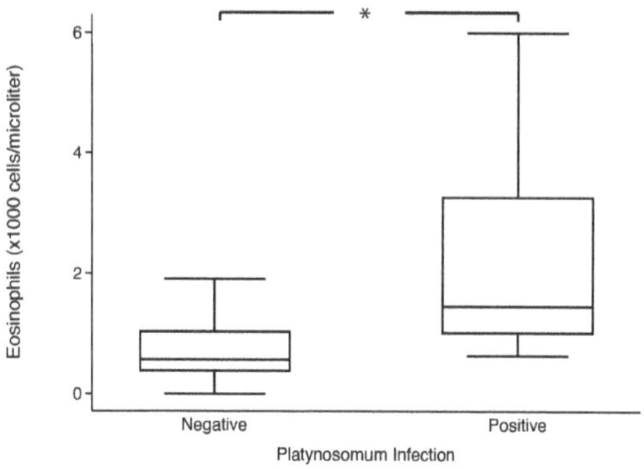

Figure 4. Box-plot comparison of alanine aminotransferase (ALT), alkaline phosphatase (ALP), and eosinophils counts between *Platinosomum*-positive and -negative cats. *, $p < 0.05$.

4. Discussion

In this study, we estimated the occurrence of and assessed factors and clinicopathological abnormalities associated with *Platynosomum* infection in cats with elevated serum ALT levels that visited the Small Animal Veterinary Teaching Hospital, Chiang Mai University, Chiang Mai, Thailand. The infection of *Platynosomum illiciens* in cats with increased ALT levels was common at a proportion of 11.63%. However, detecting this liver fluke in the fecal samples of cats can vary depending on the population tested and diagnostic techniques.

Platynosomum infection represents a significant concern in cat populations, with documented prevalence rates varying from 0.07% to 9.76% in Asia [3–10]. In Thailand, the prevalence of *Platynosomum* infection in stray cats has been reported at 0.07% (1/1485) in Bangkok, 2.67% (4/150) in Pathum Thani and Nakhon Pathom Provinces, and 2.7% (8/300) in Nakhon Nayok Province [7–9]. In contrast, the prevalence of this parasite in cats with elevated ALT levels in the current study was 11.63% (95% CI: 4.70–25.58), notably higher than previous reports in Thailand. Previous reports have revealed that *P. illiciens* infection is typically asymptomatic, with animals often harboring parasites without displaying specific clinical abnormalities [11,28]. In a recent systematic review by Silva et al., the predominant specific symptom of *Platynosomum-infected* cats was icterus (45.8%), whereas the general symptom was weakness (65.8%) [29]. In this study, 60% (3/5) of infected cats were asymptomatic, while 40% (2/5) exhibited jaundice and anorexia. Furthermore, there were no statistically significant associations between other clinical signs and the presence of *Platynosomum* infection ($p > 0.05$). These findings are consistent with several previous studies [11,28,30]. Therefore, the difference in the estimated prevalence of *Platynosomum* infection in this study compared to others in Thailand may be attributed to the fact that all cats in this study had elevated serum ALT levels, suggesting problems in the hepatobiliary system that pose a high risk for platynosomiasis. It is also important to note that the severity of clinical signs depends on several factors, including the fluke burden, the infection duration, and the animal's immune system response [14,15].

Detecting *Platynosomum* infection is challenging. The detection of *Platynosomum* eggs can vary depending on the sources of the sample and diagnostic techniques. Using fecal detection, the prevalence has been reported to be as low as 0.07%, but the prevalence of this parasite in necropsies is reportedly 3.16 to 45% [1,2]. In addition, the study of Ramos et al. revealed that microscopic egg detection from bile samples was 1.6 times more sensitive than manual trematode collection during necropsy. Regarding diagnostic techniques, many gaps remain in the history of platynosomiasis. This indicates that the infection may have occurred sporadically or has been underestimated over time due to the inefficiency of diagnostic methods [1]. However, recent studies have revealed that distinct coproparasitological evaluations, including the FLOTAC technique, using a flotation solution with a specific gravity of 1.35, and formalin–ether sedimentation, are effective tools for diagnosing *P. illiciens* [1,2]. With detection under a light microscope after these concentrating techniques, false positives are unlikely; however, a false negative can result from a too-low specific gravity of flotation solution that cannot float *Platynosomum* eggs or from excessive debris in the sample that masks the detection of the egg [16]. In one study, the formalin–ether concentration technique was superior to the sugar flotation technique with a specific gravity of 1.27 [8]. In this study, fecal flotation using a flotation solution with a specific gravity of 1.35 showed a detection rate similar to formalin–ether sedimentation. Therefore, the flotation technique with a specific gravity solution of 1.35 or formalin–ether sedimentation may be suggested for detecting trematode eggs.

Upon examination of the *Platynosomum* eggs under light microscopy, we expected golden to dark brown, elliptical, and embryonated eggs containing an operculum at one end, with a size range of 20–35 × 34–50 μm [6,31,32]. The morphology of *Platynosomum* spp. eggs found in this study was primarily consistent with previous reports. However, *Platynosomum* eggs from one cat (Cat2) were smaller (10.0–13.75 × 13.75–20.0 μm) with a thick shell, long elliptical shape, and lack of internal germ balls. We speculate that the smaller size and unusual morphology of these eggs could point to the detection of immature, atypical, unfertilized, or abnormal eggs that have been previously reported [2,32,33]. As morphological examination cannot be used for species differentiation, in this study, the species of all Platynosomum-positive eggs were identified using nucleotide sequence analyses of products from PCR assays targeting the ITS-2 region and cox-1 gene. All positive samples were 100% identical to *Platynosomum illiciens.*

Regarding age, cats' housing, and the predatory behaviors of cats, previous studies have reported a significant association with *P illiciens* infection, while sex, neutering status, and breed are not considered risk factors for platynosomiasis cases [14,15,29]. Cats older than two years have been shown to have a higher prevalence and higher odds of infection than cats aged less than 2 years [19]. Outdoor cats or free-roaming cats with hunting behaviors are at a higher risk of contracting parasitic infections, as they have increased accessibility to intermediate or paratenic hosts, which play an essential role in the life cycle of this parasite [31]. The present study demonstrated no significant associations between demographic data (age, sex, and breed), cats' housing, or predatory behavior with *P. illiciens* infection. This finding may be due to the small sample size of cats in this study. However, we found that juvenile (1–7 years) outdoor cats with predatory behavior were more frequently infected, which aligned with findings from previous studies [14,19]. The higher prevalence rates in the juvenile group could be attributed to their carnivorous habits, which are typically expressed from one year of age. The prepatent period of this parasite is 56–60 days [27,34]. Although our findings did not demonstrate a statistically significant association with the infection, it is advisable to recommend keeping cats indoors as a preventive measure to reduce the risk of exposure to intermediate or paratenic hosts of *Platynosomum* spp.

The multivariable logistic regression analysis revealed an association between deworming programs and *P. illiciens* infection. Cats with no deworming schedule were 16 times more likely to be infected with *P. illiciens* than cats with a regular schedule. This result confirmed the findings of previous studies regarding the importance of a regular deworming program in decreasing the chances of infection [35,36]. Notably, two out of five infected cats underwent a deworming regimen but still were positive for *P. illiciens* infection. Of these two cats, one received a combination of pyrantel embonate at 57.5 mg/kg combined with praziquantel at 5 mg/kg, and another one received pyrantel embonate at 5 mg/kg for over a year without a repeat deworming program. Therefore, the infection may be due to these cats not receiving appropriate deworming, which could involve using an ineffective anthelmintic agent, underdosing, or irregular deworming. Regarding the treatment efficacy of anthelmintic medications for platynosomiasis, praziquantel at 20–30 mg/kg PO once daily for three to five consecutive days, with a repeat treatment 12 weeks after the initial medication, is the most commonly used method, reported to be more efficient than fenbendazole at 50 mg/kg PO twice a day for five days [36,37]. Therefore, we recommend regular deworming with praziquantel, accompanied by at least an annual fecal examination, to reduce the risk of infection, as well as monitoring for cats over one year of age that roam freely and lack a consistent deworming program.

Regarding platynosomiasis in cats, signs of hepatocellular injury have been reported, including changes in serum ALT, with ALP usually remaining unaffected [11]. A marked increase in ALT indicates hepatic parenchymal cell damage, resulting in cytosol content leakage into the circulation. The elevation of ALP was associated with severe cholestasis, which was observed at a low frequency (16.6%) in platynosomiasis cases [28,38]. In this study, ALT and ALP showed no statistically significant differences between parasitized and non-parasitized cats. We speculate that these results were due to the small sample population of infected cats and the variability in the infectious status of the individual animals. The fluke burden and host immune status determined the severity of the parasitic infestation and the time the animals developed hepatic inflammation, leading to cholangitis or cholangiohepatitis related to clinicopathological abnormality consequences [11,28]. In this study, eosinophil counts were significantly higher in parasitized cats than in non-parasitized cats ($p = 0.014$). This result is entirely in agreement with other studies, in which eosinophilia has been reported to be more common in cats with platynosomiasis [39–41]. The study of Taylor et al. revealed that all cats experimentally infected with 125 flukes (small dose) and 1000 flukes (large dose) exhibited eosinophilia at peak level 4 to 5 months after infection [41]. It is well documented that helminths induce Th2-dominant immune responses and increase the number of eosinophils. The eosinophil chemotaxis has a direct cytotoxic effect on parasites [42].

Based on our findings, an elevation in eosinophil counts in cats with elevated serum ALT may be in accordance with *P. illiciens* infection. Nevertheless, a larger sample size is essential for a more comprehensive evaluation of the association between clinicopathologic abnormalities and the infection. Regarding other hepatobiliary enzymes, serum aspartate aminotransferase (AST) and gamma-glutamyl transferase (GGT) are interesting parameters for evaluating chronic hepatobiliary abnormalities. Therefore, further studies should investigate the changes in these enzyme parameters that might be associated with *P. illiciens* infection. Regarding the detection of fluke ova, using only coproparasitological evaluation was not robust enough to detect the infection. Fecal examination can only detect parasitic eggs during the shedding period. Thus, fecal samples should be collected and re-evaluated every 2–3 months due to the prepatent period of this fluke [28,34]. In platynosomiasis cases involving bile duct obstruction, the fluke ova cannot pass into the feces through the intestinal tract, potentially leading to misdiagnosis [11,31]. Therefore, additional diagnostic techniques, such as hepatobiliary ultrasonography and bile microscopic evaluation, should be investigated in further studies to enhance the diagnostic efficiency of *P. illiciens* infection in cats.

5. Conclusions

In conclusion, this study provided insights into the occurrence, factors, and clinicopathologic abnormalities associated with *P. illiciens* infection in cats with elevated liver enzymes. Our findings demonstrate that the presence of eosinophilia in cats, along with an increase in serum ALT levels, may suggest *P. illiciens* infection in endemic areas. These findings should raise awareness of this parasitic infection among veterinary practitioners. Furthermore, the implementation of an appropriate deworming program is strongly recommended as a preventive measure to minimize the risk of *P. illiciens* infection.

Author Contributions: Conceptualization, S.T. (Saruda Tiwananthagorn); Methodology, S.T. (Saruda Tiwananthagorn), S.T. (Sahatchai Tangtrongsup), P.C. and K.P.; Sample collection, J.P., N.C. and C.K.; Statistical analysis, S.T. (Sahatchai Tangtrongsup); Molecular analysis, S.T. (Saruda Tiwananthagorn); Formal analysis, S.T. (Saruda Tiwananthagorn), S.T. (Sahatchai Tangtrongsup), P.C., J.P. and N.C.; Data curation, S.T. (Saruda Tiwananthagorn), S.T. (Sahatchai Tangtrongsup), P.C., K.P., J.P. and N.C.; Writing—original draft preparation, P.C., S.T. (Saruda Tiwananthagorn) and S.T. (Sahatchai Tangtrongsup); Writing—review and editing, P.C., S.T. (Saruda Tiwananthagorn) and S.T. (Sahatchai Tangtrongsup). All authors have read and agreed to the published version of the manuscript.

Funding: This study was financially supported by Chiang Mai University. Grant number RG24/2556.

Institutional Review Board Statement: The research protocols were approved by the animal ethics committee of the Faculty of Veterinary Medicine, Chiang Mai University (9 July 2021; Ref. No. S19/2564) under the guidelines for the Care and Use of Experimental Animals, National Research Council of Thailand.

Informed Consent Statement: Informed consent was obtained from all subjects involved in the study.

Data Availability Statement: The sequences of *P. illiciens* generated in the current research were submitted to GenBank (DDBJ/EMBL/GenBank database Accession No. LC779845–LC779848 for ITS2 and LC779849–LC779850 for *cox1*).

Acknowledgments: We would like to express our gratitude to all the cat owners for their valuable participation and acknowledge the excellent assistance provided by Duanghatai Sripakdee at the Parasitological Laboratory of the Faculty of Veterinary Medicine, Chiang Mai University.

Conflicts of Interest: The authors declare no conflicts of interest.

References

1. Ramos, R.A.; Lima, V.F.; Monteiro, M.F.; Santana, M.d.A.; Lepold, R.; Faustino, M.A.; Rinaldi, L.; Cringoli, G.; Alves, L.C. New insights into diagnosis of *Platynosomum fastosum* (Trematoda: *Dicrocoeliidae*) in cats. *Parasitol. Res.* **2016**, *115*, 479–482. [CrossRef] [PubMed]
2. Basu, A.K.; Charles, R.A. A review of the cat liver fluke *Platynosomum fastosum* Kossack, 1910 (Trematoda: *Dicrocoeliidae*). *Vet. Parasitol.* **2014**, *200*, 1–7. [CrossRef] [PubMed]
3. Mohd-Zain, S.N.; Sahimin, N.; Pal, P.; Lewis, J.W. Macroparasite communities in stray cat populations from urban cities in Peninsular Malaysia. *Vet. Parasitol.* **2013**, *196*, 469–477. [CrossRef] [PubMed]
4. Ranaraja, D.; Mahakapuge, T.A.N.; Thilakarathne, D.S.; Dharmasuriya, J.Y.C.; Rajapakse, R. Hepatic trematode *Platynosomum sp.* (*Dicrocoeliidae*) from a domestic cat in Colombo, Sri Lanka: Case report and molecular identification. *Vet. Parasitol. Reg. Stud. Rep.* **2022**, *34*, 100762. [CrossRef] [PubMed]
5. Le, H. New findings on the cat liver flukes in some southern provinces of Vietnam. *J. Vet. Sci. Technol.* **2012**, *19*, 14–18.
6. Nguyen, H.M.; Van-Hoang, H.; Ho, L.T. *Platynosomum fastosum* (Trematoda: *Dicrocoeliidae*) from Cats in Vietnam: Morphological Redescription and Molecular Phylogenetics. *Korean J. Parasitol.* **2017**, *55*, 39–45. [CrossRef]
7. Jittapalapong, S.; Inparnkaew, T.; Pinyopanuwat, N.; Kengradomkij, C.; Sangvaranond, A.; Wongnakphet, S. Gastrointestinal parasites of stray cats in Bangkok metropolitan areas, Thailand. *Agr. Nat. Resour.* **2007**, *41*, 69–73.
8. Rojekittikhun, W.; Chaisiri, K.; Mahittikorn, A.; Pubampen, S.; Sa-Nguankiat, S.; Kusolsuk, T.; Maipanich, W.; Udonsom, R.; Mori, H. Gastrointestinal parasites of dogs and cats in a refuge in Nakhon Nayok, Thailand. *Southeast Asian J. Trop. Med. Public Health* **2014**, *45*, 31.
9. Rojekittikhun, W.; Mahittikorn, A.; Prummongkol, S.; Puangsa-art, S.; Chaisiri, K.; Kusolsuk, T. Prevalence of gastrointestinal parasitic infections in refuge dogs and cats and evaluation of two conventional examination techniques. *J. Trop. Med. Parasitol.* **2013**, *36*, 58–67.
10. Kim, J.M.; Che, J.H.; Jeong, D.H.; Kang, B.C. *Platynosomum fastosum* infestation in a domestic cat in Korea. *Lab. Anim. Res.* **2010**, *26*, 307–309. [CrossRef]
11. Carreira, V.S.; Vieira, R.F.; Machado, G.F.; Luvizotto, M.C. Feline cholangitis/cholangiohepatitis complex secondary to *Platynosomum fastosum* infection in a cat. *Rev. Bras. Parasitol. Vet.* **2008**, *17*, 184–187. [PubMed]
12. Andréia, F.C.; Gabriel, M.L.; Luciana, S.S.; Annelise, C.C.; Luis, D.G.; Paola, M.C. Resolution of a biliary obstruction caused by *Platynosomum fastosum* in a feline by a modified cholecystoduodenostomy approach–case report. *Acta Vet.* **2018**, *68*, 224–231. [CrossRef]
13. Ramos, D.G.S.; Santos, A.R.; Freitas, L.C.; Braga, I.A.; Silva, E.P.; Soares, L.M.C.; Antoniassi, N.A.B.; Furlan, F.H.; Pacheco, R.C. Feline platynosomiasis: Analysis of the association of infection levels with pathological and biochemical findings. *Rev. Bras. Parasitol. Vet.* **2017**, *26*, 54–59. [CrossRef] [PubMed]

14. Salomão, M.; Souza-Dantas, L.M.; Mendes-de-Almeida, F.; Branco, A.S.; Bastos, O.P.; Sterman, F.; Labarthe, N. Ultrasonography in hepatobiliary evaluation of domestic cats (*Felis catus*, L., 1758) infected by *Platynosomum Looss*, 1907. *Intern. J. Appl. Res. Vet. Med.* **2005**, *3*, 271–279.
15. Lima, R.L.; Pacheco, R.C.; Mendonça, A.J.; Néspoli, P.E.B.; Morita, L.H.M.; Almeida, A.; Sousa, V.R.F. *Platynosomum fastosum* in domestic cats in Cuiabá, Midwest region of Brazil. *Vet. Parasitol. Reg. Stud. Reports. Vet.* **2021**, *24*, 100582. [CrossRef] [PubMed]
16. Eisenbraun, H.; Alluwie, A.; Connolly, S.; Nair, R.; Verocai, G.G.; Ketzis, J.K. Comparison of fecal analysis methods for the detection of *Platynosomum fastosum* in naturally infected cats. *J. Feline Med. Surg.* **2020**, *22*, 395–398. [CrossRef] [PubMed]
17. Rocha, N.O.; Portela, R.W.; Camargo, S.S.; Souza, W.R.; Carvalho, G.C.; Bahiense, T.C. Comparison of two coproparasitological techniques for the detection of *Platynosomum sp.* infection in cats. *Vet. Parasitol.* **2014**, *204*, 392–395. [CrossRef] [PubMed]
18. Pinto, H.A.; Pulido-Murillo, E.A.; Braga, R.R.; Mati, V.L.T.; Melo, A.L.; Tkach, V.V. DNA sequences confirm low specificity to definitive host and wide distribution of the cat pathogen *Platynosomum illiciens* (=*P. fastosum*) (Trematoda: *Dicrocoeliidae*). *Parasitol. Res.* **2018**, *117*, 1975–1978. [CrossRef]
19. Rodriguez-Vivas, R.I.; Williams, J.J.; Quijano-Novelo, A.G.; Bolio, G.M.; Torres-Acosta, J.F. Prevalence, abundance and risk factors of liver fluke (*Platynosomum concinnum*) infection in cats in Mexico. *Vet. Rec.* **2004**, *154*, 693–694. [CrossRef]
20. Merck, Merck Manual: Veterinary Manual. 2019. Available online: https://www.merckvetmanual.com/special-subjects/reference-guides (accessed on 12 August 2023).
21. Dean, A. OpenEpi: Open Source Epidemiologic Statistics for Public Health, version 2.3.1. 2010. Available online: http://www.openepi.com (accessed on 27 September 2023).
22. Leung, T.L.F.; Donald, K.M.; Keeney, D.B.; Koehler, A.V.; Peoples, R.C.; Poulin, R. Trematode parasites of Otago Harbour (New Zealand) soft-sediment intertidal ecosystems: Life cycles, ecological roles and DNA barcodes. *New Zealand J. Mar. Freshw. Res.* **2009**, *43*, 857–865. [CrossRef]
23. Hall, T.A. BioEdit: A user-friendly biological sequence alignment editor and analysis program for Windows 95/98/NT. *Nucleic Acids Symp. Ser.* **1996**, *41*, 95–98.
24. Kumar, S.; Stecher, G.; Li, M.; Knyaz, C.; Tamura, K. MEGA X: Molecular evolutionary genetics analysis across computing platforms. *Mol. Biol. Evol.* **2018**, *35*, 1547–1549. [CrossRef]
25. Citterio, C.V.; Obber, F.; Trevisiol, K.; Dellamaria, D.; Celva, R.; Bregoli, M.; Ormelli, S.; Sgubin, S.; Bonato, P.; Rold, G.D.; et al. *Echinococcus multilocularis* and other cestodes in red foxes (*Vulpes vulpes*) of northeast Italy, 2012–2018. *Parasit. Vectors* **2021**, *14*, 29. [CrossRef] [PubMed]
26. Inoue, K.; Shishida, K.; Kawarai, S.; Takeda, S.; Minami, M.; Taira, K. Helminthes detected from wild sika deer (*Cervus nippon centralis*) in Kanto-Chubu region, Japan. *Parasitol. Intern.* **2022**, *87*, 102485. [CrossRef] [PubMed]
27. Tamura, K.; Nei, M. Estimation of the number of nucleotide substitutions in the control region of mitochondrial DNA in humans and chimpanzees. *Mol. Biol. Evol.* **1993**, *10*, 512–526.
28. Braga, R.R.; Teixeira, A.C.; Oliveira, J.A.; Cavalcanti, L.P. Prevalence of *Platynosomum fastosum* infection in free roaming cats in northeastern Brazil: Fluke burden and grading of lesions. *Vet. Parasitol.* **2016**, *227*, 20–25. [CrossRef] [PubMed]
29. Silva, W.I.; Feitosa, T.F.; Vilela, V.L.R. A systematic review and meta-analysis on the global status of *Platynosomum* sp. (Trematoda–*Dicrocoelidae*) infecting domestic cats (*Felis catus*). *Vet. Parasitol.* **2023**, *322*, 110031. [CrossRef] [PubMed]
30. Sobral, M.C.G.; Sousa, S.A.P.; Ribeiro, T.M.P.; Galvão, S.R.; Santos, R.M.; Silva, R.A.; Reis, T.S.; Dias, F.E.F.; Santos, H.D. Infection by *Platynosomum illiciens* (=*P. fastosum*) in domestic cats of Araguaína, Tocantins, northern Brazil. *Rev. Bras. Parasitol. Vet.* **2019**, *28*, 786–789. [CrossRef] [PubMed]
31. Nur-Amalina, N.; Watanabe, M.; Mazlina, M.; Che-Amat, A.; Nor-Azlina, A.A. Feline platynosomiasis in Klang Valley, Malaysia: Diagnosis, pathology and management. *Trop. Biomed.* **2022**, *39*, 26–35.
32. Soe, B.K.; Ngamniyom, A.; Adisakwattana, P.; Techangamsuwan, S.; Anuracpreeda, P.; Bowman, D.; Sukhumavasi, W. Morphology and Ultrastructural Characteristics of *Platynosomum fastosum* Kossack, 1910 Adults and Eggs. *Inter. J. Vet. Sci.* **2023**, *12*, 504–521.
33. Ferreira, A.M.R.; Almeida, E.C.P.; Labarthe, N.V. Liver fluke infection (*Platynosomum concinnum*) in Brazilian cats: Prevalence and pathology. *Feline Pract.* **1999**, *27*, 19–21.
34. Pinto, H.A.; Mati, V.L.; Melo, A.L. Experimental platynosomosis: Characterization of parasite development in the mouse model. *Vet. Parasitol.* **2015**, *211*, 40–44. [CrossRef] [PubMed]
35. Lathroum, C.N.; Shell, L.; Neuville, K.; Ketzis, J.K. Efficacy of Praziquantel in the Treatment of *Platynosomum fastosum* in Cats with Natural Infections. *Vet. Sci.* **2018**, *5*, 35. [CrossRef] [PubMed]
36. Shell, L.; Ketzis, J.; Hall, R.; Rawlins, G.P.W. Praziquantel treatment for *Platynosomum* species infection of a domestic cat on St Kitts, West Indies. *J. Feline Med. Surg. Open Rep.* **2015**, *1*, 2055116915589834. [CrossRef]
37. Daniel, A.G.; Diaz, R.F.; Camignatto, L.O.; Kage, N.K.; Pellegrino, A.; Cogliati, B. Polycystic liver disease associated with *Platynosomum fastosum* infection in a cat. *Braz. J. Vet. Pathol.* **2012**, *5*, 137–141.
38. Haney, D.R.; Christiansen, J.S.; Toll, J. Severe cholestatic liver disease secondary to liver fluke (*Platynosomum concinnum*) infection in three cats. *J. Am. Anim. Hosp. Assoc.* **2006**, *42*, 234–237. [CrossRef]
39. Xavier, F.G.; Morato, G.S.; Righi, D.A.; Maiorka, P.C.; Spinosa, H.S. Cystic liver disease related to high *Platynosomum fastosum* infection in a domestic cat. *J. Feline Med. Sur.* **2007**, *9*, 51–55. [CrossRef] [PubMed]

40. Newell, S.M.; Graham, J.P.; Roberts, G.D.; Ginn, P.E.; Greiner, E.C.; Cardwell, A.; Mauragis, D.; Knutsen, C.; Harrison, J.M.; Martin, F.G. Quantitative hepatobiliary scintigraphy in normal cats and in cats with experimental cholangiohepatitis. *Vet. Radiol. Ultrasound.* **2001**, *42*, 70–76. [CrossRef] [PubMed]
41. Taylor, D.; Perri, S. Experimental infection of cats with the liver fluke *Platynosomum concinnum*. *Am. J. Vet. Res.* **1977**, *38*, 51–54.
42. Foster, N.; Elsheikha, H.M. The immune response to parasitic helminths of veterinary importance and its potential manipulation for future vaccine control strategies. *Parasitol. Res.* **2012**, *110*, 1587–1599. [CrossRef]

animals

MDPI

Article

Current Status of *Trypanosoma grosi* and *Babesia microti* in Small Mammals in the Republic of Korea

Hyun Jung Kim, BoGyeong Han, Hee-Il Lee, Jung-Won Ju and Hyun-Il Shin *

Division of Vectors and Parasitic Diseases, Korea Disease Control and Prevention Agency, 187 Osongsaenmyeong 2-ro, Osong-eup, Heungdeok-gu, Cheongju 28159, Republic of Korea; kimhj0324@korea.kr (H.J.K.); borudd1@korea.kr (B.H.); isak@korea.kr (H.-I.L.); jupapa@korea.kr (J.-W.J.)

* Correspondence: hishin@korea.kr; Tel.: +82-43-719-8525

Simple Summary: Parasitic protozoa are a common cause of vector-borne disease outbreaks and have been identified in different groups of wild animal species. Small mammals, such as rodents and shrews, play an important role in the transmission and maintenance of parasites. To understand the risk of parasitic protozoa, it is essential to have a good understanding of the status of transmitting vectors. This study aimed to investigate the diversity and current status of parasitic protozoa, including *Trypanosoma*, *Babesia*, and *Theileria*, from small mammals in the Republic of Korea. We found that the prevalence of *Trypanosoma grosi* was 23.9% (79/331) and *Babesia microti* was 10% (33/331), while *Theileria* was not detected in small mammals. These results can be used to raise awareness of parasite infection in the Republic of Korea.

Abstract: Small mammals, such as rodents and shrews, are natural reservoir hosts of zoonotic diseases, including parasitic protozoa. To assess the risk of rodent-borne parasitic protozoa in the Republic of Korea (ROK), this study investigated the status of parasitic protozoa, namely *Trypanosoma*, *Babesia*, and *Theileria*, in small mammals. In total, 331 blood samples from small mammals were analyzed for parasites using PCR and sequenced. Samples were positive for *Trypanosoma grosi* (23.9%; $n = 79$) and *Babesia microti* (10%; $n = 33$) but not *Theileria*. Small mammals from Seogwipo-si showed the highest infection rate of *T. grosi* (48.4%), while the highest *B. microti* infection rate was observed in those from Gangneung-si (25.6%). Sequence data revealed *T. grosi* to be of the AKHA strain. Phylogenetic analysis of *B. microti* revealed the US and Kobe genotypes. *B. microti* US-type–infected small mammals were detected throughout the country, but the Kobe type was only detected in Seogwipo-si. To our knowledge, this is the first nationwide survey that confirmed *T. grosi* and *B. microti* infections at the species level in small mammals in the ROK and identified the Kobe type of *B. microti*. These results provide valuable information for further molecular epidemiological studies on these parasites.

Keywords: wild animal; rodents; shrews; protozoa; molecular epidemiology

1. Introduction

Small mammals, such as rodents and shrews, are well-known hosts and reservoirs of zoonotic diseases that pose a crucial threat to human health. Over 2000 species of rodents are distributed worldwide and live closely with humans [1,2]. Of these, approximately 200 species are reservoirs of more than 60 zoonotic diseases caused by viruses, bacteria, and parasites [3,4]. *Yersinia pestis*, *Salmonella*, and *Hantavirus* are among the most important pathogens affecting public health [5,6].

Zoonotic parasites are a common cause of vector-borne disease outbreaks and have been identified in different groups of wild animal species. Rodents play an important role in the transmission and maintenance of zoonotic parasites [6,7]. Most parasites in the blood and tissues of rodents can be transmitted to humans through contaminated food, water, rodent urine and feces, or via ectoparasites [8]. Therefore, it is necessary to investigate zoonotic parasite infections in small mammals to reduce exposure risk and predict future trends in pathogen prevalence and distribution according to seasonal and environmental changes.

Several zoonotic parasitic protozoa, such as *Trypanosoma* and *Babesia*, which cause trypanosomiasis and babesiosis, respectively, have gained importance as infectious agents. *Trypanosoma* is a parasitic hemoflagellate protozoan belonging to the Trypanosomatidae family that can infect animals and humans [9,10]. Certain species of this parasite may be transmitted by blood-feeding arthropods such as *Triatoma* sp. (kissing bug) and *Glossina* sp. (tsetse fly). *Trypanosoma* cause serious diseases in humans, such as Chagas disease (also known as American trypanosomiasis), caused by *Trypanosoma cruzi*, and African trypanosomiasis, caused by the *Trypanosoma brucei* complex [11].

Piroplasms are parasitic protozoa belonging to the genera *Babesia* and *Theileria*, which are the causative agents of babesiosis and theileriosis, respectively. Both intraerythrocytic protozoan parasites are transmitted by ticks [12,13] and are recognized for their important economic effects on the livestock industry and human health. Zoonotic babesiosis, mainly caused by *Babesia microti*, has recently posed a serious public health risk worldwide, in contrast to theileriosis, which has no zoonotic potential [14,15]. Approximately 100 *Babesia* species are known to infect animals and humans. The common clinical symptoms of babesiosis include fever, chills, fatigue, and headache [16,17]. The incidence of babesiosis has increased between 2011 and 2019 in the United States [18]. Cases of *Babesia* infections have been reported in various countries worldwide, including China, Germany, Canada, Australia, Japan, and the Republic of Korea (ROK) [19,20].

Therefore, given the importance of these three parasitic protozoa within the livestock industry and/or Public Health and as they can also be found in wild small mammals as reservoir hosts, we used molecular methods to investigate the diversity and current status of parasitic protozoa, including *Trypanosoma*, *Babesia*, and *Theileria*, in small mammals in the ROK.

2. Materials and Methods

2.1. Ethical Approval

All animal experiments were performed according to the guidelines for ethical conduct in the care and use of animals and approved by the Institutional Animal Care and Use Committee of the Korea Disease Control and Prevention Agency (approval number: KDCA-102-22). Permission to capture small mammals was obtained from each site in accordance with the Wildlife Protection and Management Act of the ROK.

2.2. Small Mammal Sampling

This study was conducted at 16 sites in the ROK in 2021 (Figure 1). Trapping was performed in the spring (March and April) and autumn (October and November). Small mammals were captured using Sherman folding live traps (BioQuip Products, Rancho Dominguez, CA, USA) baited with pieces of cheese crackers set in various habitats, including rice paddy fields, dry paddy fields, waterways, meadows, and reservoirs. Live small mammals were transferred to the laboratory and euthanized with carbon dioxide. After the identification of small mammals [2], blood was drawn from the heart using syringes, preserved in EDTA vacutainer tubes (BD, Franklin Lakes, NJ, USA), and stored at −20 °C until further use.

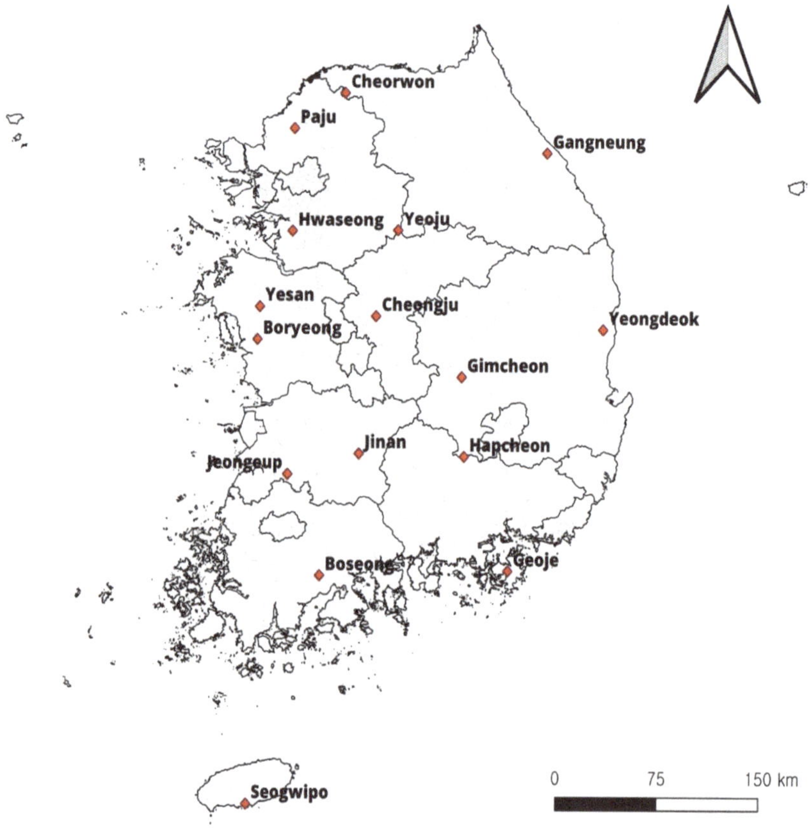

Figure 1. Trapping sites (◆) in the Republic of Korea (ROK). Map was created using the Free and Open Source QGIS (QGIS 3.28.10. Geographic Information System, http://www.qgis.org, accesed on 8 September 2023).

2.3. Molecular Detection of Parasites in Small Mammals

Genomic DNA was extracted from 331 small mammal blood samples (100 µL each) using the QIAamp DNA Blood Mini Kit, following the manufacturer's protocol (Qiagen, Hilden, Germany). To detect parasite infection, we used the extracted DNA and amplified the *ITS1* gene of *Trypanosoma* spp. using TRYP1R (5′-GGAAGCCAAGTCATCCATCG-3′) and TRYP1S (5′-CGTCCCTGCCATTTGTACACAC-′3) primer sets targeting for ~623 bp fragments [21]. PCR was performed with 20 µL using an AccuPower PCR PreMix (Bioneer, Daejeon, Republic of Korea). Each PCR mixture was composed of 1 µL of each oligonucleotide primer (10 pmol/µL), 3 µL of genomic DNA as template, and 15 µL of distilled water. The conditions were initial denaturation at 95 °C for 5 min, followed by 35 cycles of 30 s at 95 °C, 1 min at 55 °C, and 1 min at 72 °C, with a final extension at 72 °C for 10 min in a C1000^{TM} Thermal Cycler (Bio-Rad, Hercules, CA, USA). Thus, we amplified the *18S rRNA* gene of *Babesia* and *Theileria* to obtain a 561 bp fragment using the commercial AccuPower Babesia & Theileria PCR Kit (Bioneer, Daejeon, Republic of Korea) according to the manufacturer's instructions. Briefly, the lyophilized premix and primer comprised 3 µL genomic DNA as template, and 17 µL distilled water. The PCR reaction mixture was subjected to thermal cycling at 95 °C for 5 min, followed by 40 cycles of 20 s at 95 °C, and 50 s at 59 °C, with a final extension at 72 °C for 5 min. The genomic DNA of *B. microti* and *T. cruzi*, obtained from American Type Culture Collection (ATCC, Manassas, VA, USA), and

Theileria, obtained from Bioneer (Daejeon, Republic of Korea), were used as positive controls. A suitable positive control and negative control were included in each amplification reaction. Amplification products were analyzed using a QIAxcel capillary electrophoresis system (Qiagen). The DNA extraction, PCR amplification, and automated electrophoresis were performed in separate rooms to prevent contamination.

2.4. Sequencing and Phylogenetic Analysis

Positive PCR products were purified using QIAquick PCR purification kits (Qiagen) and sequenced using Sanger sequencing and an ABI PRISM 3730xl analyzer (Life Technologies, Carlsbad, CA, USA). The nucleotide sequences were compared with reference sequences obtained from GenBank using nucleotide BLAST (National Center for Biotechnology Information, NCBI). All sequences used in phylogenetic analyses were downloaded from GenBank, available through the NCBI. A phylogenetic tree was constructed using the neighbor-joining method and p-distance model in MEGA X (Pennsylvania State University, State College, PA, USA). The sequences obtained in this study were submitted to GenBank (accession numbers: OP804252, OP804253, OP804254, OP297200, and OP297201).

2.5. Statistical Analysis

Statistical analyses were performed using IBM SPSS Statistics software version 26 (IBM, Armonk, NY, USA). Pearson's chi-square test or Fisher's exact test was used to examine the association between parasitic infections in small mammals and related factors, including small mammal species, seasonal habitat type, and trapping site. $p < 0.05$ was considered significant.

3. Results

3.1. Prevalence of Parasitic Protozoa in Small Mammals in the ROK

Over the two trapping seasons in 2021, 331 small mammals belonging to eight genera and ten species were collected from 16 sites in the ROK. The results showed that 14 sites harbored parasitic protozoa, *T. grosi* or *B. microti*, infecting small mammals, with the exceptions being Boryeong-si and Cheongju-si (Table 1); *Theileria* was not detected in any of the samples. The small mammals collected from Seogwipo-si had the highest *T. grosi* infection rate at 48.4% (15/31). The results also showed that 10 sites harbored *B. microti*, with the highest *B. microti* infection rate being 32.6% (14/43) in Gangneung-si. The infection rates in Seogwipo-si and Gangneung-si were significantly higher than those in other sites ($p < 0.05$). Also, our results showed that one genotype of *B. microti* was confirmed in each site. Among positive samples of *B. microti*, the US type was found to be the dominant genotype in the ROK; only the *B. microti* Kobe type was detected in Seogwipo-si, which is the first confirmation from small mammals in the ROK (Table 1).

Table 1. Molecular prevalence of *Trypanosoma grosi* and *Babesia microti* and genetic diversity of *B. microti* in small mammals collected from different sites of the Republic of Korea (ROK) in 2021.

Site	No. Tested	No. *T. grosi*-Positive (%)	No. *B. microti*-Positive (%)	*B. microti* Genotype
Gangneung-si	43	11 (25.6)	14 (32.6) §§	US type
Hwaseong-si	33	12 (36.4)	1 (3.0)	US type
Seogwipo-si	31	15 (48.4) §	2 (6.5)	Kobe type
Jinan-gun	24	5 (20.8)	2 (8.3)	US type
Cheorwon-gun	23	6 (64.7)	1 (4.3)	US type
Geoje-si	23	6 (26.1)	0 (0.0)	
Yeoju-si	22	3 (13.6)	0 (0.0)	
Yeongdeok-gun	20	4 (20.0)	2 (10.0)	US type
Paju-si	19	6 (31.6)	6 (31.6)	US type
Jeongeup-si	18	1 (5.6)	2 (11.1)	US type
Gimcheon-si	18	1 (5.6)	0 (0.0)	

Table 1. *Cont.*

Site	No. Tested	No. *T. grosi*-Positive (%)	No. *B. microti*-Positive (%)	*B. microti* Genotype
Boseong-gun	17	1 (5.9)	1 (5.9)	US type
Hapcheon-gun	16	1 (6.3)	2 (12.5)	US type
Yesan-si	14	2 (14.3)	0 (0.0)	
Boryeong-si	8	0 (0.0)	0 (0.0)	
Cheongju-si	8	0 (0.0)	0 (0.0)	
Total	331	79 (23.9)	33 (10.0)	

§ Significantly higher *T. grosi* infection rate in Seogwipo-si vs. other sites; §§ Significantly higher *B. microti* infection rate in Gangneung-si vs. other sites.

3.2. Prevalence and Phylogenetic Analysis of T. grosi and B. microti in Small Mammals

Three species, *Apodemus agrarius* (striped field mouse), *Crocidura* sp., and *Apodemus peninsulae* (Korean field mouse), were confirmed to be infected with parasitic protozoa. *A. agrarius* was highly dominant, accounting for 87.6% (290/331) of trapped small mammals, and co-infected with *T. grosi* and *B. microti* in six individuals (Table 2).

Table 2. Prevalence of *Trypanosoma grosi* and *Babesia microti* in blood samples from small mammals collected from the ROK in 2021.

Small Mammal Species	No. Tested	No. of *T. grosi*-Positive (%)	No. of *B. microti*-Positive (%)	No. of Coinfection (%)
Apodemus agrarius	290	77 (26.5)	32 (11.0)	6 (2.0)
Crocidura sp.	23	2 (8.7)	0	0
Apodemus peninsulae	2	0	1 (50.0)	0
Craseomys regulus	4	0	0	0
Craseomys rufocanus	1	0	0	0
Cricetulus triton	1	0	0	0
Micromys minutus	6	0	0	0
Microtus fortis	2	0	0	0
Myodes regulus	1	0	0	0
Rattus norvegicus	1	0	0	0
Total	331	79 (23.9)	33 (10.0)	6 (1.8)

To detect *Trypanosoma* in small mammals, we target the *ITS1* gene of the protozoa. Of the 331 samples, 79 (23.9%) were positive and sequencing revealed 98.8% identity with *T. grosi* (AB175624), identified in two species of small mammals: *A. agrarius* (26.5%) and *Crocidura* sp. (8.7%). Of the 79 sequences identified as *T. grosi*, two representative sequences were selected without duplicates of sequences and host source. Therefore, the two *T. grosi ITS1* sequences were deposited in GenBank under the accession numbers OP297200 and OP297201. Phylogenetic analysis revealed that OP297200 (from *A. agrarius*) and OP297201 (from *Crocidura* sp.) are closely related with the previously reported *T. grosi* AKHA strain sequence (AB175624), which was isolated from *A. speciosus* in Japan (Figure 2).

A total of 33 (10.0%) samples were positive and sequencing revealed 100.0% similarity to *B. microti*, found in two species of small mammals: *A. agrarius* (11%) and *A. peninsulae* (50%). Of the 33 sequences identified as *B. microti*, three representative sequences were selected without duplicates of sequences and host source. The three *B. microti 18S rRNA* sequences were deposited in GenBank under accession numbers OP804252, OP804253, and OP804254. Phylogenetic analysis revealed that OP804252 (from *A. agrarius* and *A. peninsulae*) OP804254 (from *A. agrarius*) were closely related to *B. microti* US type (AY693840, AF231348, AB190435, and AB190459). These clusters contained *B. microti* US type that have been isolated from small mammals or humans in different regions, including USA (AY693840 and AF231348 from human) and Japan (AB190435 from *A. speciosus* and AB190459 from *M. auratus*). OP804253 (from *A. agrarius*) belonged to the *B. microti* Kobe type (KY649339 from *R. tanezumi*, AB112050 from *R. coxinga*, AB032434 from Human, AB241633 from *N. confucianus*, and KX008036 from human) (Figure 3).

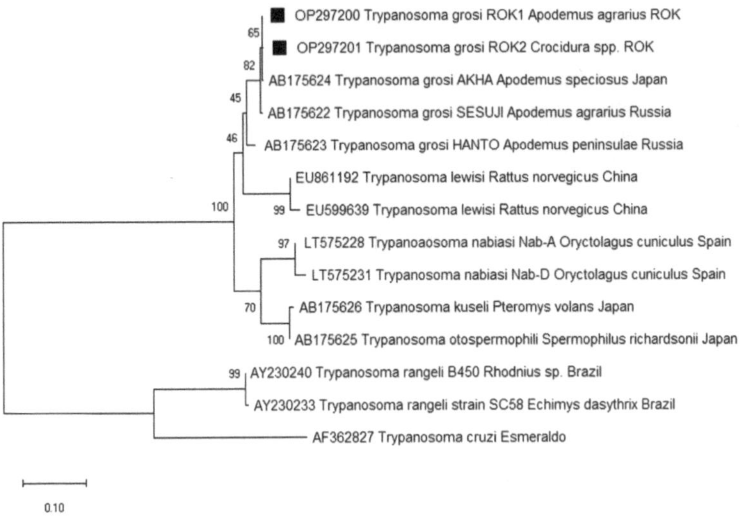

Figure 2. Phylogenetic tree based on neighbor-joining analysis of *internal transcriptional spacer 1* (*ITS1*) from GenBank and *Trypanosoma grosi*-positive small mammal specimens captured in the ROK in 2021. *Trypanosoma grosi* sequences obtained in this study are denoted by a solid square (■, GenBank accession numbers OP297200 and OP297201). The numbers on the branches indicate bootstrap percentages based on 1000 replication.

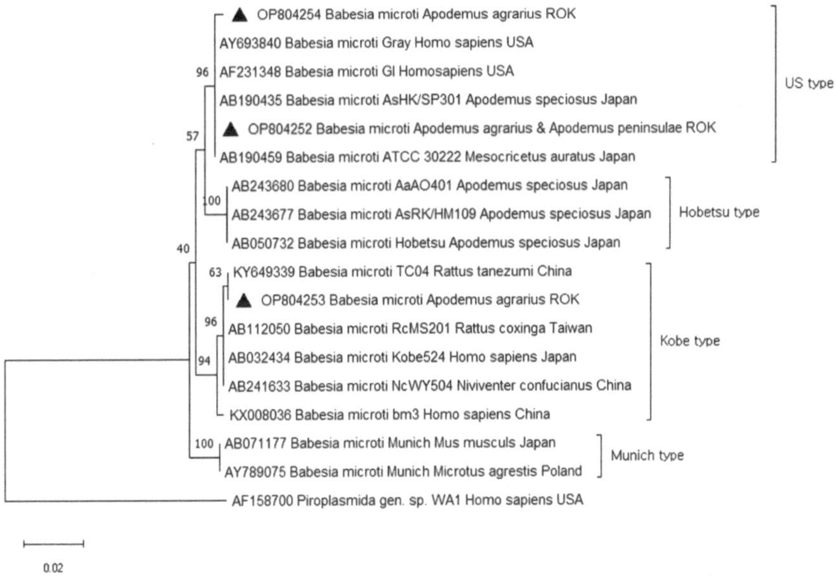

Figure 3. Phylogenetic tree based on neighbor-joining analysis of *Babesia 18S rRNA* from GenBank and *Babesia microti*-positive small mammal specimens captured in the ROK in 2021. *Babesia microti* sequences obtained in this study are denoted by a solid triangle (▲, GenBank accession numbers OP804252–OP804254). The numbers on the branches indicate bootstrap percentages based on 1000 replication.

3.3. Relationships between Season, Ecological Habitat, and T. grosi and B. microti Infections

In ecological habitats, the prevalence of *T. grosi* in small mammals from reservoirs was significantly lower than that in small mammals from other habitats (rice paddy fields, dry paddy fields, watery ways, and meadows; $p = 0.006$). The seasonal *T. grosi* positivity rate was higher in spring (26.4%; 46/174) than in autumn (21.0%; 33/157); however, the difference was not significant. The infection rate of *T. grosi* in *A. agrarius* was significantly higher than that in other species ($p = 0.0014$). The infection rates of *B. microti* per ecological habitat, season, and small mammal species are shown in Table 3. The seasonal *B. microti* positivity rate was 9.8% (17/174) for small mammals in spring and 10.2% (16/157) in autumn. There were no significant differences between seasons, small mammal species, and ecological habitats with respect to *B. microti* infection. However, *B. microti* DNA from small mammals was consistently detected throughout the study period in all habitats (Table 3).

Table 3. Relationship between season, ecological habitat, and *Trypanosoma grosi* and *Babesia microti* infections in small mammals collected from the ROK in 2021.

Categories		No. Tested	No. of *T. grosi*-Positive (%)	*p*-Value	No. of *B. microti*-Positive (%)	*p*-Value
Season	Spring	174	46 (26.4)	0.248	17 (9.8)	0.898
	Autumn	157	33 (21.0)		16 (10.2)	
Ecological habitat	Reservoir	70	8 (11.4)	0.006	9 (12.8)	0.363
	Other	261	71(27.2)		24(9.2)	
Small mammal species	*A. agrarius*	290	77 (26.5)	0.0014	32 (11.0)	0.098
	Other	41	2 (4.8)		1 (2.4)	

4. Discussion

This study was a nationwide investigation of parasitic protozoa, including *Trypanosoma*, *Babesia*, and *Theileria*, in small mammals from 16 sites across the ROK and demonstrates a wide prevalence of parasitic protozoa. Over two trapping seasons in 2021, 331 small mammals belonging to eight genera and ten species were collected from 16 sites. Blood samples from small mammals were analyzed for parasites using PCR and positive samples were sequenced. Samples were positive for *Trypanosoma grosi* (23.9%) and *Babesia microti* (10.0%) but not *Theileria*. These results suggest that the diversity of parasitic protozoa in the ROK is high and even new species can be involved in the infections.

Trypanosoma is a flagellate blood parasite found in every vertebrate class [22,23]. Over 500 species of *Trypanosoma* have been recorded worldwide [24–26]. Trypanosomes can be categorized into two groups based on their transmission route from the vector to the host [27]. The first group, known as salivarian trypanosomes, includes *T. brucei*, the causative agent of human African trypanosomiasis; this species develops in the midgut of the vector, migrates to the salivary glands or proboscis, and is transmitted to the host through the vector's saliva during the biting process. The second group, stercorarian trypanosomes, includes *T. cruzi*, the causative agent of Chagas disease. Stercorarian trypanosomes develop in the hindgut of the vector and are transmitted to the host either through the ingestion of vectors, such as fleas, or the contamination of bite wounds with vector feces [26,27]. A total of 44 *Trypanosoma* species have been identified in 144 rodent species, the majority of which belong to the stercorarian group [11,27]. Most of stercorarian trypanosomes are considered nonpathogenic, except for *T. cruzi*. To date, there has been no investigation of wild small mammals infected with *Trypanosoma* in the ROK, resulting in a critical knowledge gap.

Trypanosoma grosi belongs to the nonpathogenic stercorarian group [23,26] and has been detected in species of *Apodemus* from several countries, including Russia (*A. sylvaticus*), France (*A. sylvaticus*), Japan (*A. speciosus*), and China (*A. agrarius*) [26,28–30]. To the best of our knowledge, *T. grosi* has not yet been reported in small mammals in the ROK. Our results revealed that there were two species of small mammals (*A. agrarius* and *Crocidura* sp.) infected with *T. grosi* in the ROK, with a high infection rate of 23.9% compared to other protozoa such as *Babesia* and *Theileria*. This is also the first confirmed report of co-infection of *B. microti* and *T. grosi* in small mammals at the molecular level. To date, there has been no evidence of human infection with *T. grosi*, however, several cases have been reported of humans infected with *T. lewisi* or *T. lewisi*-like (*T. grosi*-involved group) [31–33]. Although *T. grosi* is harmless in humans, further investigation to maximize the understanding of these diseases should be considered. This study may be helpful in future molecular epidemiological studies of *T. grosi* in the ROK.

Babesia and *Theileria* are parasitic protozoa, mainly transmitted by ticks, that can infect a variety of domestic and wild animals. The prevalence of *Babesia* in small mammals was as high as 20.8% (32/154) in the ROK in 2001 [34]. However, there was a low infection rate of 0.52% (3/578) in 2008, reported for two districts (Gangwon-do and Gyeonggi-do) [35]. Both studies only detected the *B. microti* US type in small mammals. Compared to previous studies, our study showed that the infection rate of *Babesia* in small mammals in the ROK decreased from 20.8% in 2001 to 10.0% in 2021. Additionally, previous studies only confirmed the *B. microti* US type, but we used molecular methods to target *Babesia 18S rRNA* from small mammals in the ROK to identify two genetic groups of *B. microti*: the US and Kobe types. *B. microti* US-type–infected small mammals were distributed across various sites of the ROK, whereas *B. microti* Kobe-type–infected small mammals were detected only in Seogwipo-si. However, molecular evidence of *Babesia* genera in ticks has already been reported in the ROK, such as *B. duncani*, *B. venatorum*, *B. divergens*, and *B. microti* US type [36–39]. According to a recent report, the US type of *B. microti* was detected in ticks from small mammals in the ROK in 2017, while the Kobe type was only detected in ticks from the southern region of the ROK (Goheung-gun and Jeju island) [40]. This result is similar to that of our study, which showed that the Kobe type of *B. microti* was only detected in small mammals collected from Seogwipo-si, Jeju island. In the ROK, the first report of human babesiosis was reported in 1988. Around 16 cases of human babesiosis have been recorded since, most of which were imported from other countries [41,42]. Two cases were known to be locally infected with *Babesia* sp. KO-1 and *B. motasi*-like [37,43]. To date, these studies have revealed that there have been no cases of human infection by vectors transmitting *B. microti* in the ROK. Therefore, this study reinforces the importance of further studies addressing changes in *Babesia microti* genotypes in the study area.

Some species of *Theileria* such as *T. annulata* and *T. parva* are pathogenic to livestock and cause high rates of mortality in sheep and cattle [44]. Other *Theileria* spp. are thought to be less virulent, probably because of evolutionary interactions between *Theileria* and the host [45]. Recently, several studies have investigated the presence of *Theileria* spp. in various wild animals, including rodents. For example, *Theileria peramelis* was identified in *Rattus* (black rat) and *Perameles nasuta* (long-nosed bandicoots) in Australia [46]. *Theileria* sp. were successfully amplified from the rodents *Le. Striatus* and *Praomys* sp. in Gabon [47]. Those findings have not been confirmed for the zoonotic potential of *Theileria*. Therefore, further research is needed to investigate the potential consequences of infection in wildlife, especially the effect of pathogens spreading to native wildlife.

Our study has a few limitations: (1) we did not double-check positivity of *T. grosi* and/or *B. microti* in small mammals through microscopic examination. The difficulty with identifying species-level parasitic protozoa by microscopy has been described in several studies. Therefore, we used molecular methods to investigate the diversity and current status of parasitic protozoa at the species level in the small mammals. (2) only blood samples from small mammals were used to investigate the prevalence of parasitic protozoa. Further studies including the diversity and rate of parasitic infection from ectoparasites (such as fleas and ticks) in small mammals should be conducted to better understand these parasitic protozoa circulating in small mammals.

In this study, 33 of 331 samples (10.0%) were positive for *B. microti* in two small mammal species, *A. agrarius* and *A. peninsula*. These two small mammal species were confirmed to play an important role as reservoirs of *B. microti*, especially because *A. agrarius* is abundant and widely distributed across the country [48]. Although no statistical differences were observed between *B. microti* infection and season or habitat, *B. microti* DNA in small mammals was consistently detected over the study period and sites. Therefore, our findings suggest that small mammals infected with *T. grosi* and *B. microti* are widely and non-seasonally distributed throughout the ROK; however, there is no detection of *Theileria*. Additionally, this is the first study to detect *B. microti* Kobe type in small mammals in the ROK. These results provide useful information for further molecular epidemiological studies on parasites, especially those on *T. grosi* and *B. microti* transmission to humans or animals via vectors.

5. Conclusions

This study aimed to investigate the diversity and status of parasitic protozoa, including *Trypanosoma grosi*, *Babesia microti*, and *Theileria*, in small mammals in the ROK. Here, we present the rate of parasitic protozoa infection in small mammals in the study area. *T. grosi* was detected in small mammals for the first time. Furthermore, *B. microti* US-type–infected small mammals were detected throughout the country, but the Kobe type was only detected in Seogwipo-si. These results provide valuable information for further molecular epidemiological studies on these parasites.

Author Contributions: Conceptualization, H.-I.S.; methodology, H.J.K. and B.H.; writing original draft preparation, H.J.K.; writing review and editing, J.-W.J., H.-I.L. and H.-I.S. All authors have read and agreed to the published version of the manuscript.

Funding: This project was funded by the Korea Disease Control and Prevention Agency (grant number: 6331-311).

Institutional Review Board Statement: All animal experiments were performed according to the guidelines for ethical conduct in the care and use of animals and were approved by the Institutional Animal Care and Use Committee of the Korea Disease Control and Prevention Agency (approval number: KDCA-102-22).

Informed Consent Statement: Not applicable.

Data Availability Statement: The sequence data generated in the current study are available in the GenBank repository under the accession numbers OP804252, OP804253, OP804254, OP297200, and OP297201. The datasets used and/or analyzed during the present study are available upon reasonable request.

Conflicts of Interest: The authors declare no conflicts of interest.

References

1. Capizzi, D.; Bertolino, S.; Mortelliti, A. Rating the rat: Global patterns and research priorities in impacts and management of rodent pests. *Mamm. Rev.* **2014**, *44*, 148–162. [CrossRef]
2. Han, B.A.; Schmidt, J.P.; Bowden, S.E.; Drake, J.M. Rodent reservoirs of future zoonotic diseases. *Proc. Natl. Acad. Sci. USA* **2015**, *112*, 7039–7044. [CrossRef] [PubMed]
3. Morand, S.; Jittapalapong, S.; Kosoy, M. Rodents as hosts of infectious diseases: Biological and ecological characteristics. *Vector Borne Zoonotic Dis.* **2015**, *15*, 1. [CrossRef]
4. Jahan, N.A.; Lindsey, L.L.; Larsen, P.A. The role of peridomestic rodents as reservoirs for zoonotic foodborne pathogens. *Vector-Borne Zoonotic Dis.* **2021**, *21*, 133–148. [CrossRef] [PubMed]
5. Shimi, A.; Keyhani, M.; Hedayati, K. Studies on salmonellosis in the house mouse, *Mus musculus*. *Lab. Anim.* **1979**, *13*, 33–34. [CrossRef] [PubMed]
6. Meerburg, B.G. Rodents are a risk factor for the spreading of pathogens on farms. *Vet. Microbiol.* **2010**, *142*, 464–465. [CrossRef]
7. Dahmana, H.; Granjon, L.; Diagne, C.; Davoust, B.; Fenollar, F.; Mediannikov, O. Rodents as hosts of pathogens and related zoonotic disease risk. *Pathogens* **2020**, *9*, 202. [CrossRef] [PubMed]
8. Rabiee, M.H.; Mahmoudi, A.; Siahsarvie, R.; Kryštufek, B.; Mostafavi, E. Rodent-borne diseases and their public health importance in Iran. *PLoS Negl. Trop. Dis.* **2018**, *12*, e0006256. [CrossRef] [PubMed]
9. Seifollahi, Z.; Sarkari, B.; Motazedian, M.H.; Asgari, Q.; Ranjbar, M.J.; Abdolahi Khabisi, S. Protozoan parasites of rodents and their zoonotic significance in Boyer-Ahmad District, Southwestern Iran. *Vet. Med. Int.* **2016**, *2016*, 3263868. [CrossRef]
10. Tijjani, M.; Majid, R.A.; Abdullahi, S.A.; Unyah, N.Z. Detection of rodent-borne parasitic pathogens of wild rats in Serdang, Selangor, Malaysia: A potential threat to human health. *Int. J. Parasitol. Parasit. Wildl.* **2020**, *11*, 174–182. [CrossRef]

11. Pumhom, P.; Pognon, D.; Yangtara, S.; Thaprathorn, N.; Milocco, C.; Douangboupha, B.; Herder, S.; Chaval, Y.; Morand, S.; Jittapalapong, S.; et al. Molecular prevalence of *Trypanosoma* spp. in wild rodents of Southeast Asia: Influence of human settlement habitat. *Epidemiol. Infect.* **2014**, *142*, 1221–1230. [CrossRef] [PubMed]
12. Bishop, R.; Musoke, A.; Morzaria, S.; Gardner, M.; Nene, V. Theileria: Intracellular protozoan parasites of wild and domestic ruminants transmitted by ixodid ticks. *Parasitology* **2004**, *129*, S271–S283. [CrossRef] [PubMed]
13. Krause, P.J. Human babesiosis. *Int. J. Parasitol.* **2019**, *49*, 165–174. [CrossRef] [PubMed]
14. Uilenberg, G. International collaborative research: Significance of tick-borne hemoparasitic diseases to world animal health. *Vet. Parasitol.* **1995**, *57*, 19–41. [CrossRef]
15. Duh, D.; Punda-Polić, V.; Trilar, T.; Avsic-Zupanc, T. Molecular detection of *Theileria* sp. in ticks and naturally infected sheep. *Vet. Parasitol.* **2008**, *151*, 327–331. [CrossRef] [PubMed]
16. Homer, M.J.; Aguilar-Delfin, I.; Telford, S.R., 3rd; Krause, P.J.; Persing, D.H. Babesiosis. *Clin. Microbiol. Rev.* **2000**, *13*, 451–469. [CrossRef] [PubMed]
17. Krause, P.J. Babesiosis diagnosis and treatment. *Vector Borne Zoonotic Dis.* **2003**, *3*, 45–51. [CrossRef] [PubMed]
18. Swanson, M.; Pickrel, A.; Williamson, J.; Montgomery, S. Trends in Reported Babesiosis Cases—United States, 2011–2019. *MMWR Morb. Mortal Wkly. Rep.* **2023**, *72*, 273–277. [CrossRef] [PubMed]
19. Young, K.M.; Corrin, T.; Wilhelm, B.; Uhland, C.; Greig, J.; Mascarenhas, M.; Waddell, L.A. Zoonotic Babesia: A scoping review of the global evidence. *PLoS ONE* **2019**, *14*, e0226781. [CrossRef]
20. Hussain, S.; Hussain, A.; Aziz, M.U.; Song, B.; Zeb, J.; George, D.; Li, J.; Sparagano, O. A review of zoonotic babesiosis as an emerging public health threat in Asia. *Pathogens* **2022**, *11*, 23. [CrossRef]
21. Desquesnes, M.; Ravel, S.; Cuny, G. PCR identification of *Trypanosoma lewisi*, a common parasite of laboratory rats. *Kinetoplastid Biol. Dis.* **2002**, *1*, 2. [CrossRef]
22. Goodrich, I.; McKee, C.; Kosoy, M. Trypanosoma (Herpetosoma) diversity in rodents and lagomorphs of New Mexico with a focus on epizootological aspects of infection in Southern Plains woodrats (*Neotoma micropus*). *PLoS ONE* **2020**, *15*, e0244803. [CrossRef]
23. Desquesnes, M.; Gonzatti, M.; Sazmand, A.; Thévenon, S.; Bossard, G.; Boulangé, A.; Gimonneau, G.; Truc, P.; Herder, S.; Ravel, S.; et al. A review on the diagnosis of animal trypanosomoses. *Parasit. Vector.* **2022**, *15*, 64. [CrossRef]
24. Nussbaum, K.; Honek, J.; Cadmus, C.M.; Efferth, T. Trypanosomatid parasites causing neglected diseases. *Curr. Med. Chem.* **2010**, *17*, 1594–1617. [CrossRef]
25. Votýpka, J.; d'Avila-Levy, C.M.; Grellier, P.; Maslov, D.A.; Lukeš, J.; Yurchenko, V. New approaches to systematics of Trypanosomatidae: Criteria for taxonomic (re)description. *Trends Parasitol.* **2015**, *31*, 460–469. [CrossRef] [PubMed]
26. Magri, A.; Galuppi, R.; Fioravanti, M. Autochthonous *Trypanosoma* spp. in European mammals: A brief journey amongst the neglected trypanosomes. *Pathogens* **2021**, *10*, 334. [CrossRef]
27. Hoare, C.A. The trypanosomes of mammals. A zoological monograph. *Med. J. Aust.* **1972**, *1*, 140. [CrossRef]
28. Cox, F.E.G. Protozoan parasites of British small rodents. *Mamm. Rev.* **1987**, *17*, 59–66. [CrossRef]
29. Sato, H.; Osanai, A.; Kamiya, H.; Obara, Y.; Jiang, W.; Zhen, Q.; Chai, J.; Une, Y.; Ito, M. Characterization of SSU and LSU rRNA genes of three *Trypanosoma* (Herpetosoma) *grosi* isolates maintained in Mongolian jirds. *Parasitology* **2005**, *130*, 157–167. [CrossRef]
30. Guan, G.; Niu, Q.; Yang, J.; Li, Y.; Gao, J.; Luo, J.; Yin, H. *Trypanosoma* (Herpetosoma) *grosi*: First isolation from Chinese striped field mouse (*Apodemus agrarius*). *Parasitol. Int.* **2011**, *60*, 101–104. [CrossRef]
31. Howie, S.; Guy, M.; Fleming, L.; Bailey, W.; Noyes, H.; Faye, J.A.; Pepin, J.; Greenwood, B.; Whittle, H.; Molyneux, D.; et al. A Gambian infant with fever and an unexpected blood film. *PLoS Med.* **2006**, *3*, 1508–1512. [CrossRef] [PubMed]
32. Sarataphan, N.; Vongpakorn, M.; Nuansrichay, B.; Autarkool, N.; Keowkarnkah, T.; Rodtian, P.; Stich, R.W.; Jittapalapong, S. Diagnosis of a *Trypanosoma lewisi*-like (Herpetosoma) infection in a sick infant from Thailand. *J. Med. Microbiol.* **2007**, *56*, 1118–1121. [CrossRef] [PubMed]
33. Truc, P.; Büscher, P.; Cuny, G.; Gonzatti, M.I.; Jannin, J.; Joshi, P.; Juyal, P.; Lun, Z.-R.; Mattioli, R.; Pays, E.; et al. Atypical human infections by animal trypanosomes. *PLoS Negl. Trop. Dis.* **2013**, *7*, e2256. [CrossRef] [PubMed]
34. Zamoto, A.; Tsuji, M.; Wei, Q.; Cho, S.H.; Shin, E.H.; Kim, T.S.; Leonova, G.N.; Hagiwara, K.; Asakawa, M.; Kariwa, H.; et al. Epizootiologic survey for *Babesia microti* among small wild mammals in northeastern Eurasia and a geographic diversity in the beta-tubulin gene sequences. *J. Vet. Med. Sci.* **2004**, *66*, 785–792. [CrossRef] [PubMed]
35. Hong, S.H.; Lee, S.E.; Jeong, Y.I.; Kim, H.C.; Chong, S.T.; Klein, T.A.; Song, J.W.; Gu, S.H.; Cho, S.H.; Lee, W.J. Prevalence and molecular characterizations of *Toxoplasma gondii* and *Babesia microti* from small mammals captured in Gyeonggi and Gangwon Provinces, Republic of Korea. *Vet. Parasitol.* **2014**, *205*, 512–517. [CrossRef]
36. Kang, S.W.; Doan, H.T.; Choe, S.E.; Noh, J.H.; Yoo, M.S.; Reddy, K.E.; Kim, Y.H.; Kweon, C.H.; Jung, S.C.; Chang, K.Y. Molecular investigation of tick-borne pathogens in ticks from grazing cattle in Korea. *Parasitol. Int.* **2013**, *62*, 276–282. [CrossRef] [PubMed]
37. Hong, S.H.; Kim, S.Y.; Song, B.G.; Rho, J.R.; Cho, C.R.; Kim, C.N.; Um, T.H.; Kwak, Y.G.; Cho, S.H.; Lee, S.E. Detection and characterization of an emerging type of *Babesia* sp. similar to *Babesia motasi* for the first case of human babesiosis and ticks in Korea. *Emerg. Microbes Infect.* **2019**, *8*, 869–878. [CrossRef]
38. Kim, T.Y.; Kim, S.Y.; Kim, T.K.; Lee, H.I.; Cho, S.H.; Lee, W.G.; Kim, H. Molecular evidence of zoonotic *Babesia* species, other than *B. microti*, in ixodid ticks collected from small mammals in the Republic of Korea. *Vet. Med. Sci.* **2021**, *7*, 2427–2433. [CrossRef] [PubMed]

39. Bang, M.S.; Kim, C.M.; Pyun, S.H.; Kim, D.M.; Yun, N.R. Molecular investigation of tick-borne pathogens in ticks removed from tick-bitten humans in the southwestern region of the Republic of Korea. *PLoS ONE* **2021**, *16*, e0252992. [CrossRef]
40. Kim, T.Y.; Kim, S.Y.; Seo, J.; Lee, H.I.; Lee, W.G.; Kim, H. Detection of US and Kobe-type *Babesia microti* in ticks collected from small mammals of the Republic of Korea. *Entomol. Res.* **2024**, *54*, e12708. [CrossRef]
41. Kwon, H.Y.; Im, J.H.; Park, Y.K.; Durey, A.; Lee, J.S.; Baek, J.H. Two Imported Cases of Babesiosis with Complication or Co-Infection with Lyme Disease in Republic of Korea. *Korean J. Parasitol.* **2018**, *56*, 609–613. [CrossRef] [PubMed]
42. Na, Y.J.; Chai, J.Y.; Jung, B.K.; Lee, H.J.; Song, J.Y.; Je, J.H.; Seo, J.H.; Park, S.H.; Choi, J.S.; Kim, M.J. An Imported Case of Severe falciparum Malaria with Prolonged Hemolytic Anemia Clinically Mimicking a Coinfection with Babesiosis. *Korean J. Parasitol.* **2014**, *52*, 667–672. [CrossRef] [PubMed]
43. Kim, J.Y.; Cho, S.H.; Joo, H.N.; Tsuji, M.; Cho, S.R.; Park, I.J.; Chung, G.T.; Ju, J.W.; Cheun, H.I.; Lee, H.W.; et al. First Case of Human Babesiosis in Korea: Detection and Characterization of a Novel Type of *Babesia* sp. (KO1) Similar to Ovine Babesia. *J. Clin. Microbiol.* **2007**, *45*, 2084–2087. [CrossRef] [PubMed]
44. Okal, M.N.; Odhiambo, B.K.; Otieno, P.; Bargul, J.L.; Masiga, D.; Villinger, J.; Kalayou, S. *Anaplasma* and *Theileria* Pathogens in Cattle of Lambwe Valley, Kenya: A Case for Pro-Active Surveillance in the Wildlife–Livestock Interface. *Microorganisms* **2020**, *8*, 1830. [CrossRef] [PubMed]
45. Taylor, L.H.; Welburn, S.C.; Woolhouse, M.E. *Theileria annulata*: Virulence and transmission from single and mixed clone infections in cattle. *Exp. Parasitol.* **2002**, *100*, 186–195. [CrossRef] [PubMed]
46. Egan, S.L.; Taylor, C.L.; Austen, J.M.; Banks, P.B.; Northover, A.S.; Ahlstrom, L.A.; Ryan, U.M.; Irwin, P.J.; Oskam, C.L. Haemoprotozoan surveillance in peri-urban native and introduced wildlife from Australia. *Curr. Res. Parasitol. Vector-Borne Dis.* **2021**, *1*, 100052. [CrossRef]
47. Mangombi, J.B.; N'dilimabaka, N.; Lekana-Douki, J.B.; Banga, O.; Maghendji-Nzondo, S.; Bourgarel, M.; Leroy, E.; Fenollar, F.; Mediannikov, O. First investigation of pathogenic bacteria, protozoa and viruses in rodents and shrews in context of forest-savannah-urban areas interface in the city of Franceville (Gabon). *PLoS ONE* **2021**, *16*, e0248244. [CrossRef]
48. Sakka, H.; Quéré, J.P.; Kartavtseva, I.; Pavlenko, M.; Chelomina, G.; Atopkin, D.; Bogdanov, A.; Michaux, J. Comparative phylogeography of four *Apodemus* species (Mammalia: Rodentia) in the Asian Far East: Evidence of Quaternary climatic changes in their genetic structure. *Biol. J. Linn. Soc.* **2010**, *100*, 797–821. [CrossRef]

animals

MDPI

Article

First Detection of *Theileria sinensis*-like and *Anaplasma capra* in *Ixodes kashmiricus*: With Notes on *cox1*-Based Phylogenetic Position and New Locality Records

Muhammad Numan [1], Abdulaziz Alouffi [2], Mashal M. Almutairi [3], Tetsuya Tanaka [4], Haroon Ahmed [5], Haroon Akbar [6], Muhammad Imran Rashid [6], Kun-Hsien Tsai [7,*] and Abid Ali [1,*]

1 Department of Zoology, Abdul Wali Khan University Mardan, Mardan 23200, Pakistan
2 King Abdulaziz City for Science and Technology, Riyadh 12354, Saudi Arabia
3 Department of Pharmacology and Toxicology, College of Pharmacy, King Saud University, Riyadh 11451, Saudi Arabia
4 Laboratory of Infectious Diseases, Joint Faculty of Veterinary Medicine, Kagoshima University, Kagoshima 890-0065, Japan; k6199431@kadai.jp
5 Department of Biosciences, COMSATS University Islamabad (CUI), Park Road, Chak ShahZad, Islamabad 45550, Pakistan
6 Department of Parasitology, University of Veterinary and Animal Sciences, Lahore 54200, Pakistan
7 Global Health Program, Institute of Environmental and Occupational Health Sciences, College of Public Health, National Taiwan University, Taipei 106319, Taiwan
* Correspondence: kunhtsai@ntu.edu.tw (K.-H.T.); uop_ali@yahoo.com (A.A.)

Simple Summary: *Ixodes* species are the main vectors of bacteria and piroplasm for different vertebrate hosts. Research on these unexplored concerns has been neglected in different regions including Pakistan. Recently, we molecularly characterized *Ixodes kashmiricus* ticks and associated *Rickettsia* spp.; however, the *cox1* sequence and associated *Theileria* spp. and *Anaplasma* spp. for this tick are unknown. This study aimed to genetically identify *I. kashmiricus* based on the *cox1* sequence and associated *Theileria* spp. and *Anaplasma* spp. A total of 352 ticks including adult females, nymphs and males were collected from small ruminants. The BLAST results and phylogenetic analysis of the *cox1* sequence revealed a close resemblance with the *Ixodes ricinus* complex sequences. The 18S rDNA and 16S rDNA sequences showed maximum identity with *Theileria* cf. *sinensis* or *Theileria sinensis* and *Anaplasma capra*, respectively, and they phylogenetically clustered with the same species. This is the first report on the *cox1* sequence of the *I. kashmiricus* tick, new locality records, and associated *T. sinensis*-like and *A. capra*. In order to determine the epidemiology of *Ixodes* ticks and their related pathogens, a widespread tick investigation is required.

Abstract: *Ixodes* ticks transmit *Theileria* and *Anaplasma* species to a wide range of animals. The spreading of ticks and tick-borne pathogens has been attributed to transhumant herds, and research on these uninvestigated issues has been neglected in many countries, including Pakistan. Recently, we used internal transcribed spacer (ITS) and 16S ribosomal DNA partial sequences to genetically characterize *Ixodes kashmiricus* ticks and their associated *Rickettsia* spp. However, the data on its *cox1* sequence and associated *Theileria* spp. and *Anaplasma* spp. are missing. This study aimed to genetically characterize *I. kashmiricus* based on the *cox1* sequence and their associated *Theileria* spp. and *Anaplasma* spp. The *I. kashmiricus* ticks were collected from small ruminants: sheep (*Ovis aries*) and goats (*Capra hircus*) of transhumant herds in district Shangla, Dir Upper and Chitral, Khyber Pakhtunkhwa (KP), Pakistan. Out of 129 examined hosts, 94 (72.87%) (56 sheep and 38 goats) were infested by 352 ticks, including adult females (175; 49.7%) followed by nymphs (115; 32.7%) and males (62; 17.6%). For molecular analyses, 121 ticks were subjected to DNA isolation and PCR for the amplification of the *cox1* sequence for *I. kashmiricus*, 18S rDNA for *Theileria* spp. and 16S rDNA sequences for *Anaplasma* spp. The obtained *cox1* sequence showed 89.29%, 88.78%, and 88.71% identity with *Ixodes scapularis*, *Ixodes gibbosus*, and *Ixodes apronophorus*, respectively. Phylogenetically, the present *cox1* sequence clustered with the *Ixodes ricinus* complex. Additionally, the 18S rDNA sequence showed 98.11% maximum identity with *Theileria* cf. *sinensis* and 97.99% identity with

Theileria sinensis. Phylogenetically, *Theileria* spp. clustered with the *T.* cf. *sinensis* and *T. sinensis*. In the case of *Anaplasma* spp., the 16S rDNA sequence showed 100% identity with *Anaplasma capra* and phylogenetically clustered with the *A. capra*. PCR-based DNA detection targeting the amplification of *groEL* and *flaB* sequences of *Coxiella* spp. and *Borrelia* spp., respectively, was unsuccessful. This is the first phylogenetic report based on *cox1* and new locality records of *I. kashmiricus*, and the associated *T. sinensis*-like and *A. capra*. Significant tick surveillance studies are needed in order to determine the epidemiology of *Ixodes* ticks and their associated pathogens.

Keywords: *Ixodes kashmiricus*; *cox1*; *Theileria sinensis*-like; *Anaplasma capra*; transhumant herds; Pakistan

1. Introduction

Genus *Ixodes* (Acari: Ixodidae: Prostriata) developed during the Mesozoic era's cretaceous period (65–95 million years ago) [1,2]. The *Ixodes* genus comprises more than 265 species, which are divided based on morphology into 18 subgenera [3]. Among them, the largest subgenus *Ixodes* comprises 18 species and includes the most studied ticks [4]. *Ixodes* ticks are known to adopt in particular environmental conditions for survival and development, and these are considered to limit their dispersal [3,5]. Climatic conditions and the availability of a suitable host are the two most important factors determining the distribution and abundance of *Ixodes* ticks. *Ixodes* ticks have been commonly found in woodland or mixed forest and grassland, which provide moist vegetation and approximately 80% humidity—a critical threshold for the survival and development of these ticks [2,5].

Ixodes ticks are known to parasitize a wide range of hosts including birds, reptiles, and mammals [3]. These ticks are capable of transmitting pathogens of medical and veterinary importance like *Theileria* spp., *Anaplasma* spp., *Coxiella* spp., *Babesia* spp. and *Borrelia* spp. [5–8]. Hard ticks, particularly of the *Haemaphysalis, Dermacentor, Ixodes* and *Rhipicephalus* genera are the primary vectors that transmit *Anaplasma* spp. [9,10]. To date, only two species of *Anaplasma* spp. like *Anaplasma phagocytophilum* have been detected in *Ixodes* ticks such as *Ixodes ricinus* [11], *Ixodes trianguliceps* [12], *Ixodes scapularis* [13] and *Ixodes frontalis* [14], while the *Anaplasma capra* has been detected in *Ixodes persulcatus* [15]. Several other hard tick species, including *Haemaphysalis longicornis, Haemaphysalis qinghaiensis, Rhipicephalus sanguineus, Rhipicephalus turinicus, Rhipicephalus haemaphysaloides, Rhipicephalus microplus* and *Dermacentor everstianus* have been shown as carrier of *A. capra* [9,10,16]. Similarly, some piroplasm species such as *Theileria annae* in *Ixodes hexagonus* [17], *Theileria fuliginosa* in *Ixodes australiensis* [18], and *Theileria* spp. in *I. ricinus* [14] have been described. On the other hand, *Ha. qinghaiensis* is the only known vector of *Theileria sinensis* [19]. *Coxiella* spp. such as *Coxiella burnetii* [7], and *Borrelia* spp. such as *Borrelia burgdorferi, Borrelia miyamotoi, Borrelia genospecies* and "*Candidatus* Borrelia sibirica" of the relapsing fever group, have been detected in the *Ixodes* ticks [6,20].

The identification of ticks, particularly those belonging to the genus *Ixodes*, has been traditionally based on morphological features, such as the shape of the spiracular plates, grooves of the scutum and punctations [2,4,21]. However, these methods are often insufficient for accurate identification and differentiation, particularly for *Ixodes* and other closely related ticks [22,23]. Molecular techniques have been alternatively used for the accurate identification and differentiation of different tick species [22,24–28]. Some genetic markers, such as *cox1*, 16S ribosomal DNA (rDNA) and internal transcribed spacer (ITS), have been shown suitable for the accurate delineation of ticks [21,29–32]. *Ixodes kashmiricus* tick has been reported based on ITS and 16S rDNA sequences, and their associated *Rickettsia* spp. has been reported based on *gltA* and *ompA* sequences [21]. However, genetic data based on *cox1* sequence for *I. kashmiricus* and associated *Theileria* spp. and *Anaplasma* spp. are missing. Herein, *I. kashmiricus* ticks were for the first time genetically characterized based on a mitochondrial *cox1* sequence and screened for associated *Theileria* spp. and *Anaplasma* spp. in Khyber Pakhtunkhwa (KP), Pakistan.

2. Materials and Methods

2.1. Ethical Approval

This study was approved by the Advance Studies Research Board (ASRB: Dir/A&R/AWKUM/2022/9396) committee members of Abdul Wali Khan University, Mardan KP, Pakistan. The oral permission was obtained from the owners of the transhumant herds during the host's observation and tick collection.

2.2. Study Area and Tick Collection

This study was conducted in district Shangla (34°46′34.6″ N 72°40′45.8″ E), Dir Upper (35°13′23.4″ N 71°55′12.2″ E) and Chitral (35°50′11.7″ N 71°48′18.0″ E) of KP, Pakistan. These districts are highly mountainous, with an elevation approximately 3000–3500 m (m), and situated in the north or northwest of KP. The elevation study map was designed in ArcGIS 10.3.1, using the "Global Positioning System" to determine the locations of the collection sites (Figure 1). Tick specimens were collected from small ruminants in transhumant herds during May–July 2022 in district Shangla, Dir Upper and Chitral. The ticks were isolated from the host body carefully via tweezers to avoid any external damage to the specimens. The tick specimens were washed in distilled water followed by 70% ethanol and preserved in 100% ethanol in 1.5 mL Eppendorf tubes for further experiments.

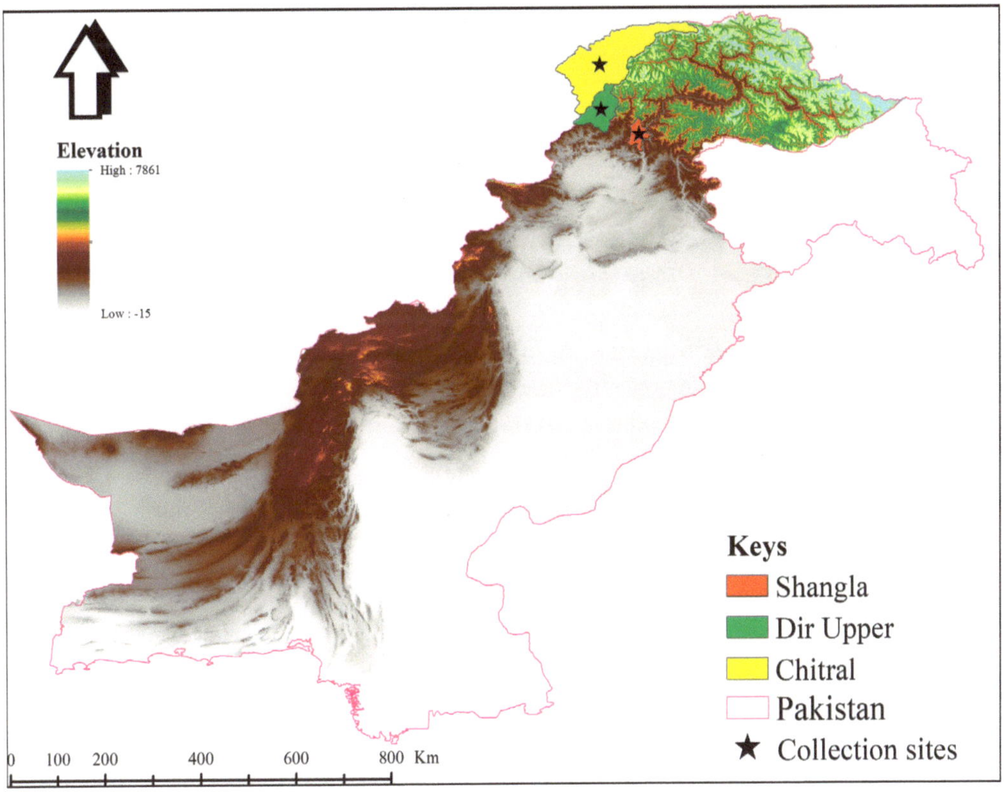

Figure 1. Map showing the locations (black stars) where *Ixodes* ticks were collected during this study.

2.3. Morphological Identification of Ticks

The collected tick specimens were morphologically identified under a stereozoom microscope (StereoBlue-euromex SB.1302-1, Arnhem, The Netherlands) using standard morphological identification keys [2,21,33].

2.4. DNA Isolation and PCR

Individually, 121 ticks including 20 males, 44 adult females and 27 nymphs from sheep, as well as 16 females and 14 nymphs from goats, were selected and subjected to molecular analyses. Before the DNA isolation, tick specimens were washed with distilled water followed by 70% ethanol and kept in an incubator (30–45 min) until dried. The specimens were cut with sterile scissors and homogenized in 200–300 µL phosphate-buffered saline (pH 7.4, 137 mM NaCl, 2.7 mM KCl, 8 mM Na_2HPO_4, 2 mM KH_2PO_4) using a micro-pestle. The genomic DNA was extracted using a phenol–chloroform protocol [34], and the isolated DNA pellet was diluted by the addition of 20–30 µL of "nuclease-free" PCR water. The isolated genomic DNA was quantified via NanoDrop (Nano-Q, Optizen, Daejeon, Republic of Korea) and stored at −20 °C.

The tick genomic DNA of *I. kashmiricus* (1 male, 2 adult females, and 2 nymphs) were subjected to conventional PCR (GE-96G, BIOER, Hangzhou, China) for the amplification of mitochondrial cytochrome C oxidase 1 (*cox1*) sequence. Each PCR reaction mixture contained 25 µL volume—comprising 1 µL of each primer (10 µM), 2 µL of template DNA (50–100 ng/µL), 8.5 µL of PCR water "nuclease-free" and 12.5 µL of Dream*Taq* green MasterMix (2×) (Thermo Scientific, Waltham, MA, USA).

All extracted genomic DNA was used for the screening of associated pathogens based on genetic markers such as 18S rRNA for *Theileria* spp., 16S rRNA for *Anaplasma* spp., *groEL* for *Coxiella* spp. and *flaB* for *Borrelia* spp. Each PCR contained a positive control (DNA of *Anaplasma marginale, Theileria annulata, Coxiella burnetii* and *Borrelia anserina* for pathogens and genomic DNA of *Hy. anatolicum* for ticks) and a negative control ("nuclease-free" PCR water instead of DNA). The primers used in this study and their thermocycler conditions are given in Table 1.

Table 1. List of the primers used to amplify target DNA of the *Ixodes kashmiricus* and associated *Theileria* and *Anaplasma* species.

Organism/Marker	Primer Sequences 5′-3′	Amplicons	Annealing Temperature	References
Ticks/*cox1*	HC02198: TAAACTTCAGGGTGACCAAAAAATCA	649 bp	55 °C	[35]
	LCO1490: GGTCAACAAATCATAAAGATATTGG			
Anaplasma spp./16S rDNA	EHR16SD: GGTACCYACAGAAGAAGTCC	344 bp	55 °C	[36]
	EHR16SR: TAGCACTCATCGTTTACAGC			
Theileria spp./18S rRNA	18S_F: GGTAATTCTAGAGCTAATACATGAGC	897 bp	56 °C	This study
	18S_R: ACAATAAAGTAAAAAACAYTTCAAAG			
Coxiella spp./*groEL* *	CoxGrF1: TTTGAAAAYATGGGCGCKCAAATGGT	619 bp	56 °C	[37]
	CoxGrR2: CGRTCRCCAAARCCAGGTGC			
	CoxGrF2: GAAGTGGCTTCGCRTACWTCAGACG			
	CoxGrFR1: CCAAARCCAGGTGCTTTYAC			
Borrelia spp./*flab*	Fla SS: AAGAGCTGAAGAGCTTGGAATG	354 bp	55 °C	[38]
	Fla RS: CTTTGATCACTTATCATTCTAATAGC			

* Nested PCR.

The PCR-amplified products were electrophoresed on a 1.5% agarose gel and visualized under ultraviolet light in the Gel Documentation System (BioDoc-It™ Imaging Systems UVP, LLC, Upland, CA, USA). PCR-positive samples were purified by using a DNA Clean & Concentrator Kit (Zymo Research, Irvine, CA, USA) by following the manufacturer's instructions.

2.5. DNA Sequencing and Phylogenetic Analysis

All amplified amplicons of *cox1* (5: 1 male, 2 adult females, and 2 nymphs) for ticks, 18S rDNA (2: 1 adult female and 1 nymph) for *Theileria* spp. and 16S rDNA (4: 2 adult females and 2 nymphs) for *Anaplasma* spp. were sequenced (Macrogen Inc., Seoul, Republic of Korea) by Sanger sequencing. The obtained sequences were trimmed/edited via SeqMan v. 5 (DNASTAR, Inc., Madison/WI, USA) for the removal of poor reading sequences and subjected to Basic Local Alignment Search Tool (BLAST, https://blast.ncbi.nlm.nih.gov/Blast.cgi, accessed on: 10 July 2022) at the National Center for Biotechnology Information (NCBI, https://www.ncbi.nlm.nih.gov/, accessed on: 10 July 2022). After BLAST, maximum identity sequences of the most similar/subgenus species were downloaded in FASTA format from the NCBI. Obtained sequences were aligned with the downloaded sequences using ClustalW multiple alignments in BioEdit Sequence Alignment Editor v. 7.0.5 [39]. The phylogenies were constructed individually for each DNA sequence of tick and associated pathogens through the Maximum Likelihood statistical method and Kimura 2-parameter model in Molecular Evolutionary Genetics Analysis (MEGA-X) with a bootstrapping value of 1000 [40]. The coding sequences like *cox1* sequences were aligned by using MUSCLE algorithms [41].

2.6. Statistical Analyses

All recorded data such as the numbers of collected ticks and their life stages in the three districts, as well as associated pathogens like *Theileria* spp. and *Anaplasma* spp., were arranged in the spreadsheet (Microsoft Excel v. 2016, Microsoft 365®) for descriptive statistical analyses. The differences were considered significant at a *p*-value less than 0.05 under chi-square tests using the GraphPad Prism v. 8 (Inc., San Diego, CA, USA).

3. Results

3.1. Morphological Identification and Description of Ixodes kashmiricus

Altogether, 352 *I. kashmiricus* ticks (Table 2) were collected in this study and morphologically identified. During this study, 94 out of 129 (72.87%) hosts of small ruminants including 56 sheep and 38 goats were infested by 352 ticks comprising adult females (175/352, 49.7%) followed by nymphs (115/352, 32.7%) and males (62/352, 17.6%). A significantly high prevalence of *I. kashmiricus* was found on sheep (271/352, 77%) followed by goats (81/352, 23%) in transhumant herds.

Furthermore, other tick species were not found co-infesting sheep and goats afflicted by *I. kashmiricus* ticks. During collection from district Chitral, only an adult female of *I. kashmiricus* was found on sheep. Details of host records, prevalence of ticks, and detection of *Theileria* and *Anaplasma* species in the selected districts are summarized in Table 2.

3.2. Sequences and Phylogenetic Relationship of Ticks

A sum of five ticks' (one male, two adult females and two nymphs) genomic DNA was amplified via PCR targeting the *cox1* sequence. The BLAST analysis of the *cox1* sequence of *I. kashmiricus* showed 89.29% maximum identity with *I. scapularis* followed by 88.78% with *Ixodes gibbosus* and 88.71% with *Ixodes apronophorus* from Canada, Turkey and Russia, respectively. The obtained 16S rDNA sequence for *I. kashmiricus* was identical to the sequences of the same species from Pakistan (MW578839). Therefore, the 16S rDNA sequence was not included in further analysis. The obtained *cox1* sequence of *I. kashmiricus* was submitted to GenBank under the accession number OR244356.

Phylogenetically, the *cox1* sequence was clustered to the species of the subgenus *Ixodes ricinus* complex such as *I. apronophorus* (MH784873) reported from Russia. Furthermore, the *cox1* sequence formed sister clades with *I. ricinus* complex such as *I. scapularis, I. gibbosus, Ixodes acuminatus, Ixodes redikorzevi, Ixodes laguri, Ixodes inopinatus, Ixodes ricinus,* and *Ixodes affinis* reported from Canada, Turkey, Malta, Romania, Serbia, Tunisia, Italy and the United States (Figure 2).

Table 2. Prevalence of identified *Ixodes kashmiricus* ticks and their life stages and molecular detection of associated *Theileria* spp. and *Anaplasma* spp.

Location/ Districts	Host	Infested/ Observed Hosts	Numbers of Ticks/Life Stages			Total Collected Ticks	*p* Value	Number of Ticks Subjected to DNA Isolation	Amplified *cox1* for *Ixodes kashmiricus*	Amplified 18S rDNA for *Theileria*	Amplified 16S rDNA for *Anaplasma*
			Males	Adult Females	Nymphs						
Shangla	Sheep	42/47	37	123	72	232	<0.001	77 (17M, 38F, 22N)	2 (1M, 1F)	1 (1F)	2 (1F, 1N)
	Goats	27/39	11	24	25	60		25 (13F, 12N)	1 (1F)	1 (1N)	1 (1N)
Dir Upper	Sheep	13/18	9	16	13	38		14 (3M, 6F, 5N)	1 (1N)	0	1 (1F)
	Goats	11/16	5	11	5	21		5 (3F, 2N)	1 (1N)	0	0
Chitral	Sheep	1/5	0	1	0	1		0	0	0	0
	Goats	0/4	0	0	0	0		0	0	0	0
Total Sheep (%)		56/70 (80)	46	140	85	271 (77)		91 (20M, 44F, 27N)	3 (1M, 1F, 1N)	1 (1F)	3 (2F, 1N)
Total Goats (%)		38/59 (64.4)	16	35	30	81 (23)		30 (16F, 14N)	2 (1F, 1N)	1 (1N)	1 (1N)
Overall Total (%)		94/129 (72.87)	62 (17.6)	175 (49.7)	115 (32.7)	352 (100)		121 (20M, 60F, 41N)	5 (1M, 2F, 2N)	2 (1.65) (1F, 1N)	4 (3.3) (2F, 2N)

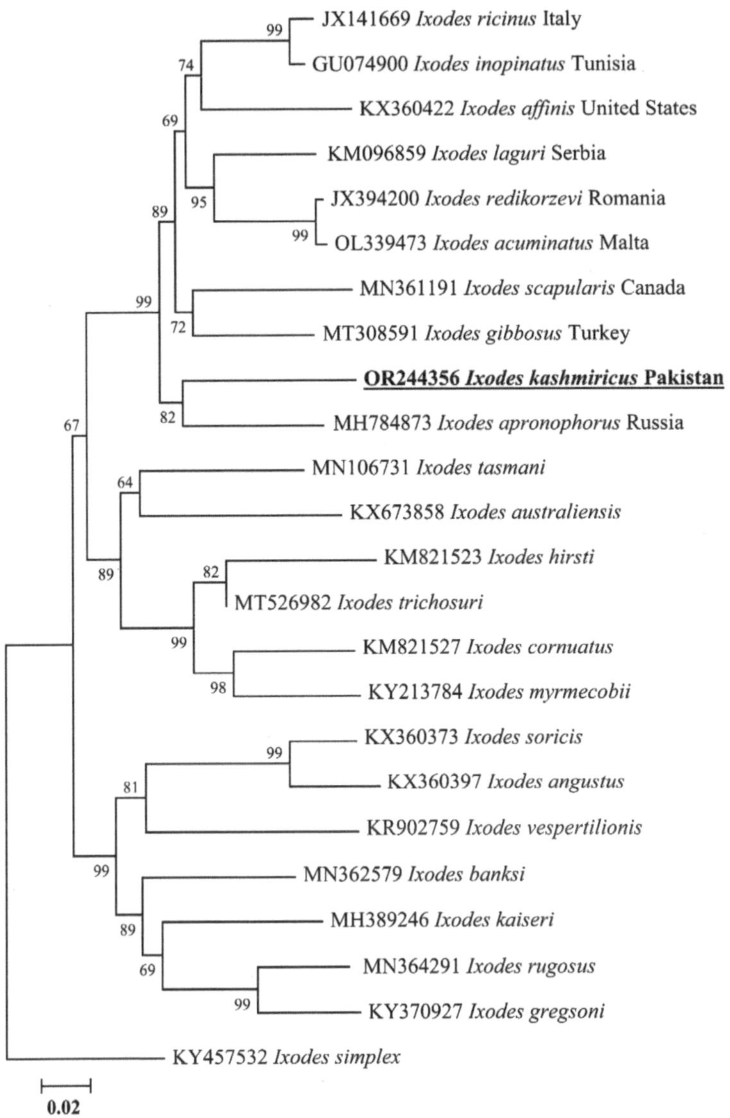

Figure 2. Phylogenetic tree of *Ixodes* species based on the *cox1* sequences. The *cox1* sequence of *Ixodes simplex* belonging to the subgenus *Eschatocephalus* was taken as an outgroup. The obtained *cox1* sequence was highlighted with bold and underlined fonts (OR244356).

3.3. Sequences and Phylogenetic Relationship of Theileria spp. and Anaplasma spp.

Among all molecularly analyzed ticks, *Theileria* spp. and *Anaplasma* spp. DNA were detected in two (1.65%: one adult female and one nymph) and four (3.3%: two adult females and two nymphs) *I. kashmiricus* ticks, respectively (Table 2). Moreover, other pathogens such as *Coxiella* spp. and *Borrelia* spp. based on *groEL* and *flaB* markers, respectively, were not amplified by PCR.

The 18S rDNA sequence of *Theileria* spp. showed 98.11% maximum identity with *Theileria* cf. *sinensis* reported from South Africa, which was followed by 97.99–97.87% identity with *T. sinensis* reported from Malaysia and China. Similarly, the 16S rDNA sequence of *Anaplasma* spp. showed 100% identity with *A. capra* reported from the Republic of Korea, China, and Iraq. The obtained 18S rDNA sequence of *T. sinensis*-like and 16S rDNA sequence of *A. capra* were submitted to GenBank (OR244360: *T. sinensis*-like and OR244358: *A. capra*). The details regarding the detection rate of *T. sinensis*-like and *Anaplasma capra* are shown in Table 2.

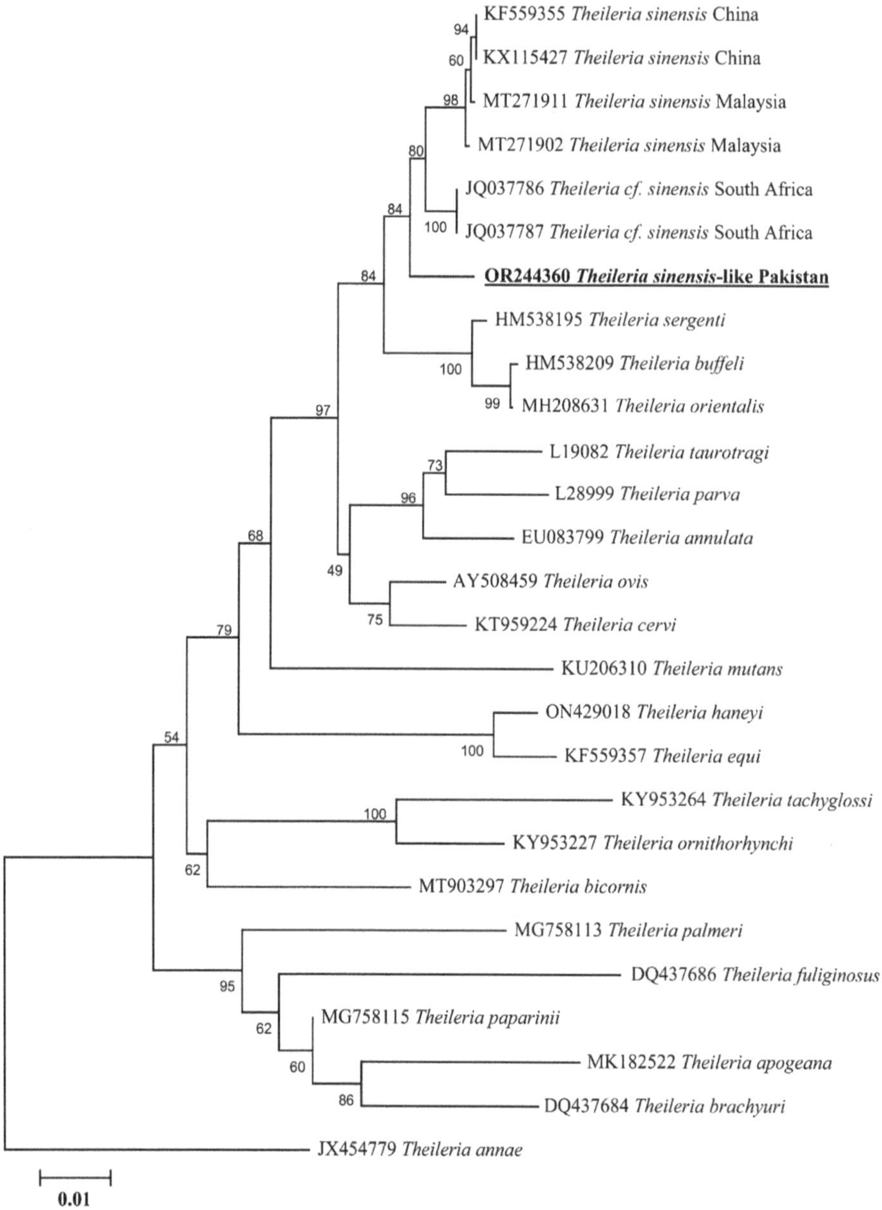

Figure 3. Phylogenetic tree of *Theileria* species based on the 18S rDNA sequences. The 18S rDNA sequence of *Theileria annae* was taken as an outgroup. The obtained 18S rDNA sequence was highlighted with bold and underlined fonts (OR244360).

The phylogenetic tree of the 18S rDNA sequence for *T. sinensis*-like clustered with *T. sinensis* (JQ037786-JQ037787) reported from South Africa and *T.* cf. *sinensis* reported from Malaysia (MT271902 and MT271911) and China (KX115427 and KF559355). It formed a sister clade with the sequences of *Theileria sergenti, Theileria buffeli* and *Theileria orientalis* (Figure 3). In the case of 16S rDNA, *A. capra* clustered to the corresponding species reported from South Korea (LC432114), China (MG869594), and Iraq (ON872236) (Figure 4).

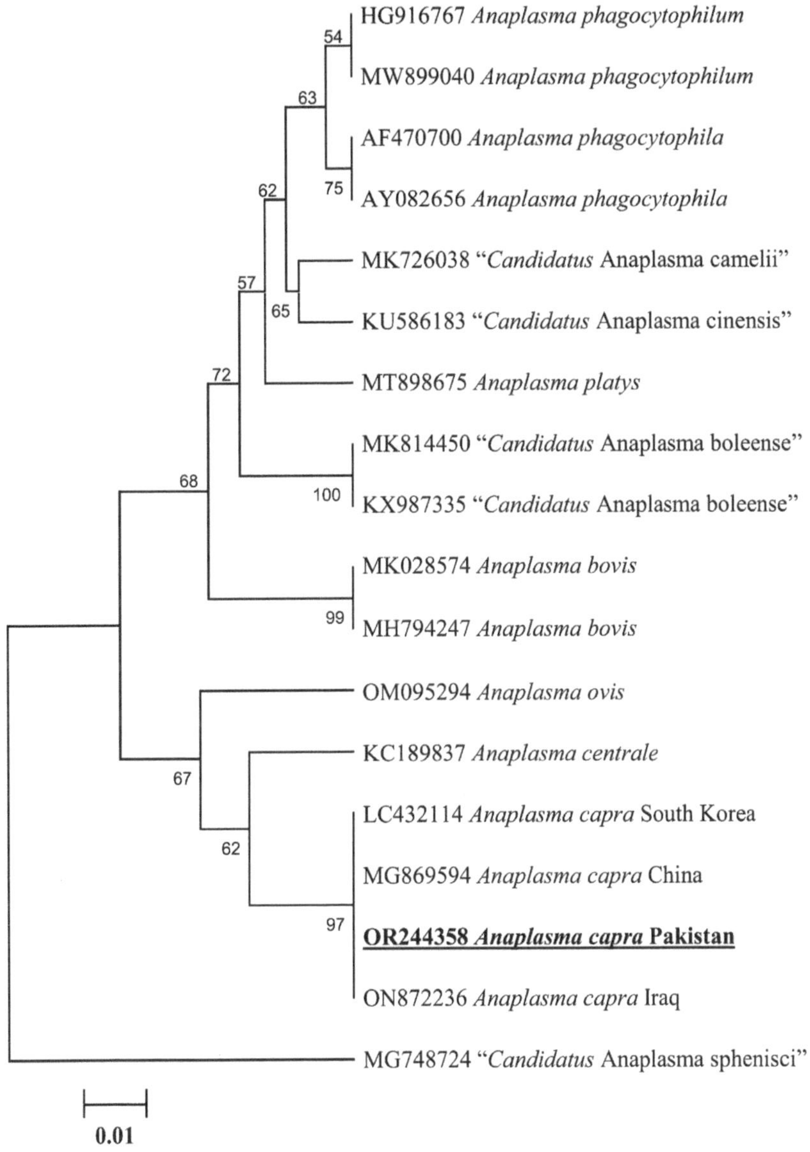

Figure 4. Phylogenetic tree of *Anaplasma* species based on the 16S rDNA sequences. The 16S rDNA sequence of "*Candidatus* Anaplasma sphenisci" was taken as an outgroup. The obtained 16S rDNA sequence was highlighted with bold and underlined fonts (OR244358).

4. Discussions

Ixodes ticks are known to transmit various pathogens such as *Anaplasma* spp., *Theileria* spp., *Coxiella* spp., *Rickettsia* spp., and *Borrelia* spp. to different hosts [5,7,11,14,20,21,42,43]. Genetic data of *I. kashmiricus* based on *cox1* sequence and their associated pathogens like *Theileria* spp. and *Anaplasma* spp. are missing. To date, a total of five *Ixodes* spp. such as *Ixodes hyatti* (Peshawar) [44], *Ixodes redikorzevi* (Kaghan) [45], *Ixodes stromi* [46], *I. kashmiricus* (Kashmir and Shangla) [21,33] and an undetermined *Ixodes* spp. (Swat) [47] have been reported in Pakistan. Among these, only *I. kashmiricus* (Shangla) has been characterized based on the morphology and molecular level [21]. In addition to the *Rickettsia* spp., *I. kashmiricus* associated with any other pathogens have not been characterized. In the current study, *I. kashmiricus* ticks collected from sheep and goats in district Shangla, Dir Upper and Chitral were characterized based on the *cox1* sequence and their associated *T. sinensis*-like and *A. capra* for the first time.

Small ruminants such as sheep and goats were found infested by *I. kashmiricus* ticks. The significantly higher infestation of *I. kashmiricus* on sheep among small ruminants shows that this tick prefers sheep as a host. The majority of the *Ixodes* spp. in the *I. ricinus* complex are associated with small ruminants: sheep and goats [33,48], which graze in areas having favorable climate conditions [49]. Similarly, the study districts are mountainous, having temperatures in the winter season below 10 °C, in the summer season 15–30 °C, a high relative humidity of ~70–80%, and precipitation throughout the year approximately 1000–1400 mm (climate-data.org, [26]). Notably, these transhumant herds seasonally migrate toward the frontiers of the country in northern and northwest areas during the spring and summer seasons (March–September). The frontiers of the country lie in the Palearctic region, which has a high prevalence and distribution of *Ixodes* ticks because of the availability of suitable environmental conditions [21,33,50]. Moreover, this transhumant movement of the infested hosts can enhance the dispersal of the *I. kashmiricus* ticks and associated pathogens to novel localities [51].

Ixodes kashmiricus ticks have been described previously by Pomerantzev [33] in India and then genetically characterized by Numan et al. [21] in Pakistan. These locations are at approximately 300 km (km) distance, while the current study's new localities, Dir Upper and Chitral, are about ~130 km and ~165 km away from the previous collection site in the district Shangla, respectively. These ticks were collected from highly mountainous areas (up to 3000–3500 m elevation), as other members of the *I. ricinus* complex have been reported from hilly ranges in the Palearctic and Oriental regions [50]. Until the present study, only ITS and 16S rDNA sequences are freely available for *I. kashmiricus* in GenBank. Herein, we provided for the first time a cox1 sequence for *I. kashmiricus*, which shared a high identity with the *I. ricinus* complex. The morphological compatibility of *I. kashmiricus* was confirmed by molecular characterization, as the 16S rDNA sequence has close resemblance to the *I. ricinus* complex and clustered with the same species, which was previously reported in Pakistan. Whereas, due to the unavailability of *cox1* sequences for *I. kashmiricus* in GenBank, the obtained *cox1* sequence clustered to the *I. ricinus* complex from the Neotropical and Palearctic regions. The topologies of the constructed phylogeny for *I. kashmiricus* were paralleled to the sequences of *I. ricinus* complex [4,21].

Until the present study, except for the undetermined *Rickettsia* spp., no other pathogens in *I. kashmiricus* have been reported [21]. Ticks of the *I. ricinus* complex are the main vector of piroplasmids such as *T. annae*, *T. fuliginosa*, and undetermined *Theileria* spp. [14,17,18]. These *Ixodes*-associated pathogens have been genetically characterized based on the 18S rDNA. This genetic marker has been demonstrated to be valuable for determining evolutionary studies of protozoans [52–55]. The suggested identity or threshold level of 18S rDNA locus for *Theileria* spp. to be considered the same species is 99.3% [52]. However, the use of various parameters to determine genetic distances has led to insufficient use of this measure [18,55]. For instance, *Theileria fuliginosa* [18] and *Theileria ornithorhynchi* [56] have been considered similar species with 97.6% and 98.2% maximum identity, respectively. Likewise, the corresponding sequence of *T. sinensis*-like detected in *I. kashmiricus* showed

98.11% maximum identity. Furthermore, the obtained 18S rDNA sequence of *T. sinensis*-like showed a minimum nucleotide difference of 16 bp with the sequences of *T. sinensis*, which showed 1.89% (16/844 bp) genetic difference. Due to the high genetic differences, this species was considered as *T. sinensis*-like or related to *T. sinensis*. Similarly, the phylogeny of the obtained 18S rDNA sequence indicated a similar relationship or related to the *T. sinensis* reported from the Old World. The constructed phylogeny work has a comparable topology to those demonstrated by Loh et al. [18], whereas *Theileria* spp. is derived from *Ixodes* spp.

Ixodes ticks collected from cattle and sheep have been reported as a vector of *A. phagocytophilum* and *A. capra* [11–15]. This study presents the first evidence for *A. capra* in *I. kashmiricus*. For the molecular identification of *Anaplasma* spp., a highly conserved 16S rDNA sequence has been used historically [27,57]. Likewise, the 16S rDNA sequence of *A. capra* was detected in *I. kashmiricus*, which was reported for the first time. The present study reported *T. sinensis*-like and *A. capra* in *I. kashmiricus* ticks infesting small ruminants that closely related to the corresponding species. The zoonotic pathogenicity of the *T. sinensis*-like and *A. capra* was detected in this survey remains to be examined considering the significance of piroplasm and bacterial species as an agent of novel emerging infectious agents carried by *I. kashmiricus* ticks.

5. Conclusions

A new locality for *I. kashmiricus* was recorded, and its phylogenetic position based on the *cox1* sequence was delineated for the first time. Based on a phylogenetic analysis, the *I. kashmiricus* tick is closely related and clustered with the species of same subgenus—the *I. ricinus* complex. *Theileria sinensis*-like and *A. capra* were detected in *I. kashmiricus* for the first time. These findings may help to further understand the epidemiology of the *I. kashmiricus* tick and its associated *Theileria* and *Anaplasma* species, and they may strengthen the need for tick and tick-borne pathogen surveillance programs.

Author Contributions: A.A. (Abid Ali) designed the study. M.M.A., T.T., A.A. (Abdulaziz Alouffi), H.A. (Haroon Ahmed), M.I.R. and K.-H.T., carried out the experiments, wrote the initial draft and H.A. (Haroon Ahmed) and H.A. (Haroon Akbar) critically reviewed the final draft. A.A. (Abid Ali) and M.N. performed the phylogenetic and statistical analysis. All authors have read and agreed to the published version of the manuscript.

Funding: The researchers supporting project number (RSP2023R494), King Saud University, Riyadh, Saudi Arabia. This research was also partially funded by the National Science and Technology Council, grant number: 112-2327-B-002-008.

Institutional Review Board Statement: This study was approved by the Advance Studies Research Board (ASRB: Dir/A&R/AWKUM/2022/9396) committee members of Abdul Wali Khan University, Mardan KP, Pakistan.

Informed Consent Statement: Not applicable.

Data Availability Statement: All the relevant data are within the manuscript.

Acknowledgments: The authors acknowledge the financial support provided by the Higher Education Commission (HEC) and Pakistan Science Foundation (PSF). The researchers supporting project number (RSP2023R494), King Saud University, Riyadh, Saudi Arabia.

Conflicts of Interest: The authors declare no conflict of interest.

References

1. Clifford, C.M.; Sonenshine, D.E.; Keirans, J.E.; Kohls, G.M. Systematics of the subfamily Ixodinae (Acarina: Ixodidae). 1. the subgenera of *Ixodes*. *Ann. Entomol. Soc. Am.* **1973**, *66*, 489–500. [CrossRef]
2. Filippova, N.A. Ixodid ticks of the subfamily Ixodinae. Fauna SSSR, Paukoobraznye, Izdatielstwo Nauka, Moskwa–Leningrad. *Arachnida* **1977**, *4*, 272–283. (In Russian)
3. Guglielmone, A.A.; Petney, T.N.; Robbins, R.G. Ixodidae (Acari: Ixodoidea): Descriptions and redescriptions of all known species from 1758 to December 31, 2019. *Zootaxa* **2020**, *4871*, 1–322.

4. Rar, V.; Yakimenko, V.; Tikunov, A.; Vinarskaya, N.; Tancev, A.; Babkin, I.; Epikhina, T.; Tikunova, N. Genetic and morphological characterization of *Ixodes apronophorus* from Western Siberia, Russia. *Ticks Tick Borne Dis.* **2020**, *11*, 101284. [CrossRef] [PubMed]
5. Pietzsch, M.E.; Medlock, J.M.; Jones, L.; Avenell, D.; Abbott, J.; Harding, P.; Leach, S. Distribution of *Ixodes ricinus* in the British Isles: Investigation of historical records. *Med. Vet. Entomol.* **2005**, *19*, 306–314. [PubMed]
6. Lommano, E.; Bertaiola, L.; Dupasquier, C.; Gern, L. Infections and coinfections of questing *Ixodes ricinus* ticks by emerging zoonotic pathogens in Western Switzerland. *Appl. Environ. Microbiol.* **2012**, *78*, 4606–4612. [CrossRef]
7. Pilloux, L.; Baumgartner, A.; Jaton, K.; Lienhard, R.; Ackermann-Gäumann, R.; Beuret, C.; Greub, G. Prevalence of *Anaplasma phagocytophilum* and *Coxiella burnetii* in *Ixodes ricinus* ticks in Switzerland: An underestimated epidemiologic risk. *New Microbes New Infect.* **2019**, *27*, 22–26. [CrossRef]
8. Sands, B.; Lihou, K.; Lait, P.; Wall, R. Prevalence of *Babesia* spp. pathogens in the ticks *Dermacentor reticulatus* and *Ixodes ricinus* in the UK. *Acta Trop.* **2022**, *236*, 106692. [CrossRef]
9. Guo, W.P.; Zhang, B.; Wang, Y.H.; Xu, G.; Wang, X.; Ni, X.; Zhou, E.M. Molecular identification and characterization of *Anaplasma capra* and *Anaplasma platys*-like in *Rhipicephalus microplus* in Ankang, Northwest China. *BMC Infect. Dis.* **2019**, *19*, 1–9.
10. Altay, K.; Erol, U.; Sahin, O.F. The first molecular detection of *Anaplasma capra* in domestic ruminants in the central part of Turkey, with genetic diversity and genotyping of *Anaplasma capra*. *Trop. Anim. Health Prod.* **2022**, *54*, 129. [CrossRef]
11. Aktas, M.; Vatansever, Z.; Altay, K.; Aydin, M.F.; Dumanli, N. Molecular evidence for *Anaplasma phagocytophilum* in *Ixodes ricinus* from Turkey. *Trans. R. Soc. Trop. Med. Hyg.* **2010**, *104*, 10–15. [PubMed]
12. Bown, K.J.; Begon, M.; Bennett, M.; Woldehiwet, Z.; Ogden, N.H. Seasonal dynamics of *Anaplasma phagocytophila* in a rodent-tick (*Ixodes trianguliceps*) system, United Kingdom. *Emerg. Infect. Dis.* **2003**, *9*, 63. [CrossRef] [PubMed]
13. Ogden, N.H.; Lindsay, L.R.; Hanincová, K.; Barker, I.K.; Bigras-Poulin, M.; Charron, D.F.; Heagy, A.; Francis, C.M.; O'Callaghan, C.J.; Schwartz, I.; et al. Role of migratory birds in introduction and range expansion of *Ixodes scapularis* ticks and of *Borrelia burgdorferi* and *Anaplasma phagocytophilum* in Canada. *Appl. Environ. Microbiol.* **2008**, *74*, 1780–1790.
14. Remesar, S.; Diaz, P.; Prieto, A.; García-Dios, D.; Panadero, R.; Fernández, G.; Brianti, E.; Díez-Baños, P.; Morrondo, P.; López, C.M. Molecular detection and identification of piroplasms (*Babesia* spp.; *Theileria* spp.) and *Anaplasma phagocytophilum* in questing ticks from northwest Spain. *Med. Vet. Entomol.* **2021**, *35*, 51–58. [PubMed]
15. Li, H.; Zheng, Y.C.; Ma, L.; Jia, N.; Jiang, B.G.; Jiang, R.R.; Huo, Q.B.; Wang, Y.W.; Liu, H.B.; Chu, Y.L.; et al. Human infection with a novel tick-borne *Anaplasma* species in China: A surveillance study. *Lancet Infect. Dis.* **2015**, *15*, 663–670. [CrossRef] [PubMed]
16. Ali, A.; Ullah, S.; Numan, M.; Almutairi, M.; Alouffi, A.; Tanaka, T. First report on tick-borne pathogens detected in ticks infesting stray dogs near butcher shops. *Front. Vet. Sci.* **2023**, *10*, 1246871.
17. Camacho, A.T.; Pallas, E.; Gestal, J.J.; Guitián, F.J.; Olmeda, A.S.; Telford III, S.R.; Spielman, A. *Ixodes hexagonus* is the main candidate as vector of *Theileria annae* in northwest Spain. *Vet. Parasitol.* **2003**, *112*, 157–163. [CrossRef]
18. Loh, S.M.; Paparini, A.; Ryan, U.; Irwin, P.; Oskam, C. Identification of Theileria fuliginosa-like species in *Ixodes australiensis* ticks from western grey kangaroos (*Macropus fuliginosus*) in Western Australia. *Ticks Tick Borne Dis.* **2018**, *9*, 632–637.
19. Wang, Y.; Wang, B.; Zhang, Q.; Li, Y.; Yang, Z.; Han, S.; Yuan, G.; Wang, S.; He, H. The common occurrence of *Theileria ovis* in Tibetan Sheep and the first report of *Theileria sinensis* in yaks from southern Qinghai, China. *Acta Parasitol.* **2021**, *66*, 1177–1185.
20. Sabitova, Y.; Rar, V.; Tikunov, A.; Yakimenko, V.; Korallo-Vinarskaya, N.; Livanova, N.; Tikunova, N. Detection and genetic characterization of a putative novel *Borrelia genospecies* in *Ixodes apronophorus*/*Ixodes persulcatus*/*Ixodes trianguliceps* sympatric areas in Western Siberia. *Ticks Tick Borne Dis.* **2023**, *14*, 102075. [CrossRef]
21. Numan, M.; Islam, N.; Adnan, M.; Zaman Safi, S.; Chitimia-Dobler, L.; Labruna, M.B.; Ali, A. First genetic report of *Ixodes kashmiricus* and associated *Rickettsia* sp. *Parasit. Vectors* **2022**, *15*, 378. [CrossRef] [PubMed]
22. Barker, S.C.; Walker, A.R.; Campelo, D. A list of the 70 species of Australian ticks; diagnostic guides to and species accounts of *Ixodes holocyclus* (paralysis tick), *Ixodes cornuatus* (southern paralysis tick) and *Rhipicephalus australis* (Australian cattle tick); and consideration of the place of Australia in the evolution of ticks with comments on four controversial ideas. *Int. J. Parasitol.* **2014**, *44*, 941–953. [PubMed]
23. Cheng, T.Y.; Chen, Z.; Li, Z.B.; Liu, G.H. First report of *Ixodes nipponensis* infection in goats in China. *Vector-Borne Zoonotic Dis.* **2018**, *18*, 575–578. [CrossRef] [PubMed]
24. Lv, J.; Wu, S.; Zhang, Y.; Chen, Y.; Feng, C.; Yuan, X.; Jia, G.; Deng, J.; Wang, C.; Wang, Q.; et al. Assessment of four DNA fragments (*COI*, 16S rDNA, ITS2, 12S rDNA) for species identification of the Ixodida (Acari: Ixodida). *Parasit. Vectors* **2014**, *7*, 93.
25. Ali, A.; Zahid, H.; Zeb, I.; Tufail, M.; Khan, S.; Haroon, M.; Bilal, M.; Hussain, M.; Alouffi, A.S.; Muñoz-Leal, S. Risk factors associated with tick infestations on equids in Khyber Pakhtunkhwa, Pakistan, with notes on *Rickettsia massiliae* detection. *Parasit. Vectors* **2021**, *14*, 363. [CrossRef]
26. Ali, A.; Numan, M.; Khan, M.; Aiman, O.; Muñoz-Leal, S.; Chitimia-Dobler, L.; Labruna, M.B.; Nijhof, A.M. *Ornithodoros* (*Pavlovskyella*) ticks associated with a *Rickettsia* sp. in Pakistan. *Parasit. Vectors* **2022**, *15*, 138.
27. Khan, Z.; Shehla, S.; Alouffi, A.; Kashif Obaid, M.; Zeb Khan, A.; Almutairi, M.M.; Numan, M.; Aiman, O.; Alam, S.; Ullah, S.; et al. Molecular survey and genetic characterization of *Anaplasma marginale* in ticks collected from livestock hosts in Pakistan. *Animals* **2022**, *12*, 1708. [CrossRef]
28. Ahmad, I.; Ullah, S.; Alouffi, A.; Almutairi, M.M.; Khan, M.; Numan, M.; Safi, S.Z.; Chitimia-Dobler, L.; Tanaka, T.; Ali, A. Description of Male, Redescription of Female, Host Record, and Phylogenetic Position of *Haemaphysalis danieli*. *Pathogens* **2022**, *11*, 1495. [CrossRef]

29. Mangold, A.J.; Bargues, M.D.; Mas-Coma, S. Mitochondrial 16S rDNA sequences and phylogenetic relationships of species of *Rhipicephalus* and other tick genera among Metastriata (Acari: Ixodidae). *Parasitol. Res.* **1998**, *84*, 478–484. [CrossRef]
30. Kovalev, S.Y.; Golovljova, I.V.; Mukhacheva, T.A. Natural hybridization between *Ixodes ricinus* and *Ixodes persulcatus* ticks evidenced by molecular genetics methods. *Ticks Tick Borne Dis.* **2016**, *7*, 113–118. [CrossRef]
31. Karim, S.; Budachetri, K.; Mukherjee, N.; Williams, J.; Kausar, A.; Hassan, M.J.; Adamson, S.; Dowd, S.E.; Apanskevich, D.; Arijo, A.; et al. A study of ticks and tick-borne livestock pathogens in Pakistan. *PLoS Negl. Trop. Dis.* **2017**, *11*, e0005681. [CrossRef] [PubMed]
32. Khan, S.M.; Khan, M.; Alouffi, A.; Almutairi, M.M.; Numan, M.; Ullah, S.; Obaid, M.K.; Islam, Z.U.; Ahmed, H.; Tanaka, T.; et al. Phylogenetic Position of *Haemaphysalis kashmirensis* and *Haemaphysalis cornupunctata*, with Notes on *Rickettsia* spp. *Genes* **2023**, *14*, 360. [CrossRef] [PubMed]
33. Pomerantzev, B.I. New ticks of the family Ixodidae. *Parazitol Sborn Zool Inst Akad Nauk. SSSR* **1948**, *10*, 20–24. (In Russian)
34. Sambrook, J.; Fritsch, E.F.; Maniatis, T. *Molecular Cloning: A Laboratory Manual*, 2nd ed.; Cold Spring Harbor Laboratory Press: Cold Spring Harbor, NY, USA, 1989.
35. Folmer, O.; Hoeh, W.R.; Black, M.B.; Vrijenhoek, R.C. DNA primers for amplification of mitochondrial *cytochrome c oxidase* subunit I from diverse metazoan invertebrates. *Mol. Mar. Biol. Biotech.* **1994**, *3*, 294–299.
36. Inokuma, H.; Raoult, D.; Brouqui, P. Detection of *Ehrlichia platys* DNA in brown dog ticks (*Rhipicephalus sanguineus*) in Okinawa Island, Japan. *J. Clin. Microbiol.* **2000**, *38*, 4219–4221. [CrossRef]
37. Duron, O.; Noël, V.; McCoy, K.D.; Bonazzi, M.; Sidi-Boumedine, K.; Morel, O.; Vavre, F.; Zenner, L.; Jourdain, E.; Durand, P.; et al. The recent evolution of a maternally-inherited endosymbiont of ticks led to the emergence of the Q fever pathogen, *Coxiella burnetii*. *PLoS Pathog.* **2015**, *11*, e1004892. [CrossRef]
38. Stromdahl, E.Y.; Williamson, P.C.; Kollars, T.M., Jr.; Evans, S.R.; Barry, R.K.; Vince, M.A.; Dobbs, N.A. Evidence of *Borrelia lonestari* DNA in *Amblyomma americanum* (Acari: Ixodidae) removed from humans. *J. Clin. Microbiol.* **2003**, *41*, 5557–5562. [CrossRef]
39. Hall, T.; Biosciences, I.; Carlsbad, C.J.G.B.B. BioEdit: An important software for molecular biology. *GERF Bull Biosci.* **2011**, *2*, 60–61.
40. Kumar, S.; Stecher, G.; Li, M.; Knyaz, C.; Tamura, K. MEGA X: Molecular evolutionary genetics analysis across computing platforms. *Mol. Biol. Evol.* **2018**, *35*, 1547. [CrossRef]
41. Edgar, R.C. MUSCLE: Multiple sequence alignment with high accuracy and high throughput. *Nucleic Acids Res.* **2004**, *32*, 1792–1797. [CrossRef]
42. Schouls, L.M.; Van De Pol, I.; Rijpkema, S.G.; Schot, C.S. Detection and identification of *Ehrlichia*, *Borrelia burgdorferi* sensu lato, and *Bartonella* species in Dutch *Ixodes ricinus* ticks. *J. Clin. Microbiol.* **1999**, *37*, 2215–2222. [CrossRef] [PubMed]
43. Parola, P.; Paddock, C.D.; Socolovschi, C.; Labruna, M.B.; Mediannikov, O.; Kernif, T.; Abdad, M.Y.; Stenos, J.; Bitam, I.; Fournier, P.E.; et al. Update on tick-borne rickettsioses around the world: A geographic approach. *Clin. Microbiol. Rev.* **2013**, *26*, 657–702. [CrossRef] [PubMed]
44. Clifford, C.M.; Hoogstraal, H.; Kohls, G.M. *Ixodes hyatti*, n. sp.; and *I. shahi*, n. sp. (Acarina: Ixodidae), parasites of pikas (Lagomorpha: Ochotonidae) in the Himalayas of Nepal and West Pakistan. *J. Med. Entomol.* **1971**, *8*, 430–438. [CrossRef] [PubMed]
45. Begum, F.; Wisseman, C.L., Jr.; Casals, J. Tick-borne viruses of west Pakistan: II. Hazara virus, a new agent isolated from *Ixodes redikorzevi* ticks from the Kaghan valley, w. Pakistan. *Am. J. Epidemiol.* **1970**, *92*, 192–194. [CrossRef]
46. Filippova, N.A. *Ixodes stromi*, new species of tick and its systematic position in Ixodinae. *Zool. Zhurnal.* **1957**, *36*, 864–869. (In Russian)
47. Begum, F.; Wisseman, C.L., Jr.; Traub, R. Tick-borne viruses of West Pakistan: I. Isolation and general characteristics. *Am. J. Epidemiol.* **1970**, *92*, 180–191. [CrossRef]
48. Hoogstraal, H. The epidemiology of tick-borne Crimean-Congo hemorrhagic fever in Asia, Europe, and Africa. *J. Med. Entomol.* **1979**, *15*, 307–417. [CrossRef]
49. Gilbert, L.; Aungier, J.; Tomkins, J.L. Climate of origin affects tick (*Ixodes ricinus*) host-seeking behavior in response to temperature: Implications for resilience to climate change? *Ecol. Evol.* **2014**, *4*, 1186–1198. [CrossRef]
50. Filippova, N.A. Forms of sympatry and possible ways of microevolution of closely related species of the group *Ixodes ricinus-persulcatus* (Ixodidae). *Acta Zool. Litu.* **2002**, *12*, 215–227. [CrossRef]
51. Volkova, V.V.; Howey, R.; Savill, N.J.; Woolhouse, M.E. Sheep movement networks and the transmission of infectious diseases. *PLoS ONE* **2010**, *5*, e11185. [CrossRef]
52. Schnittger, L.; Yin, H.; Gubbels, M.J.; Beyer, D.; Niemann, S.; Jongejan, F.; Ahmed, J.S. Phylogeny of sheep and goat *Theileria* and *Babesia* parasites. *Parasitol. Res.* **2003**, *91*, 398–406. [PubMed]
53. Schnittger, L.; Rodriguez, A.E.; Florin-Christensen, M.; Morrison, D.A. *Babesia*: A world emerging. *Infect. Genet. Evol.* **2012**, *12*, 1788–1809. [CrossRef] [PubMed]
54. Sivakumar, T.; Hayashida, K.; Sugimoto, C.; Yokoyama, N. Evolution and genetic diversity of *Theileria*. *Infect. Genet. Evol.* **2014**, *27*, 250–263. [CrossRef]
55. Ullah, K.; Numan, M.; Alouffi, A.; Almutairi, M.M.; Zahid, H.; Khan, M.; Islam, Z.U.; Kamil, A.; Safi, S.Z.; Ahmed, H.; et al. Molecular Characterization and assessment of risk factors associated with *Theileria annulata* infection. *Microorganisms* **2022**, *10*, 1614. [CrossRef] [PubMed]

56. Paparini, A.; Macgregor, J.; Ryan, U.M.; Irwin, P.J. First molecular characterization of *Theileria ornithorhynchi* Mackerras, 1959: Yet another challenge to the systematics of the piroplasms. *Protist* **2015**, *166*, 609–620. [CrossRef] [PubMed]
57. Alam, S.; Khan, M.; Alouffi, A.; Almutairi, M.M.; Ullah, S.; Numan, M.; Islam, N.; Khan, Z.; Aiman, O.; Zaman Safi, S.; et al. Spatio-temporal patterns of ticks and molecular survey of *Anaplasma marginale*, with notes on their phylogeny. *Microorganisms* **2022**, *10*, 1663. [CrossRef]

animals MDPI

Article

Sarcocystis spp. Macrocysts Infection in Wildfowl Species in Eastern Baltic Region: Trends in Prevalence in 2011–2022

Petras Prakas [1,*], Jolanta Stankevičiūtė [2], Saulius Švažas [1], Evelina Juozaitytė-Ngugu [1], Dalius Butkauskas [1] and Rasa Vaitkevičiūtė-Balčė [2]

1 Nature Research Centre, Akademijos 2, 08412 Vilnius, Lithuania; saulius.svazas@gamtc.lt (S.Š.); evelina.ngugu@gamtc.lt (E.J.-N.); dalius.butkauskas@gamtc.lt (D.B.)
2 Agriculture Academy, Vytautas Magnus University, Studentų Str 11, Kaunas District, 53361 Akademija, Lithuania; jolanta.stankeviciute1@vdu.lt (J.S.); rasa.vaitkeviciute1@vdu.lt (R.V.-B.)
* Correspondence: prakaspetras@gmail.com; Tel.: +370-684-07151

Simple Summary: Macrocysts of the protozoan *Sarcocystis rileyi*, resembling rice grains, are found in the muscles of numerous duck species. Meat from ducks contaminated with *S. rileyi* is not suitable for consumption, and severe *S. rileyi* infections are harmful to infected birds. This *Sarcocystis* species was reported only in North America for a long time. However, recently, the number of reports of *S. rileyi* infection in European wild ducks has increased. The present research is the first long-term surveillance study on macrocysts in wildfowl species. Overall, 3268 individuals of wildfowl species collected in Lithuania, Latvia, Russia, and Belarus were tested for macrocysts of *Sarcocystis*. The examined parasite was detected in 237 Mallards and in two Eurasian Teals. Macrocysts isolated from 37 Mallards were examined by DNA methods and identified as *S. rileyi*. The infection rates of macrocysts in Mallards in the examined region ranged from 3.1% to 8.7%. The fluctuations in infection rates were observed in different years, months, and studied countries. Based on the results of the current research and previous studies, the prevalence of macrocysts depends on the duck species, age of the bird, year, geographical region, and hunting season but is not related to the sex of the bird.

Abstract: Wildfowl meat infected with *S. rileyi* macrocysts is not suitable for human consumption. Ducks are among the main game birds in Europe, and *S. rileyi* infections cause significant economic losses. In 2011–2022, a total of 2649 anseriforms collected in Lithuania and 619 Mallards (*Anas platyrhynchos*) hunted in the Kaliningrad region of Russia, Belarus, and Latvia were tested for macrocysts. In Lithuania, macrocysts were detected in 206 of 2362 Mallards (8.7%) and in two of 88 (2.3%) Eurasian Teals (*Anas crecca*). The prevalence of macrocysts in the other three countries, Belarus (5.9%), Russia (5.0%), and Latvia (3.1%), was similar. For species identification, macrocysts isolated from 37 Mallards (21 from Lithuania, 8 from Russia, 6 from Belarus, and 2 from Latvia) were subjected to sequencing of the *ITS1* region. Based on DNA analysis, *S. rileyi* was confirmed in all tested birds. By comparing the infection rates of macrocysts in Mallards in Lithuania, significant differences were observed in different years (p = 0.036), and a significantly higher prevalence of infection was established in November–December than in September–October (p = 0.028). Given the amount of data per decade on the prevalence of *S. rileyi*, awareness of infection needs to be increased.

Keywords: ducks; hunting; macrocysts; *Sarcocystis rileyi*; infection rates; long-term monitoring

1. Introduction

The genus *Sarcocystis* is a large group of worldwide distributed protozoan parasites that infect mammals, birds, and reptiles. These parasites have an obligatory two-host life cycle [1]. The intermediate host gets infected through food or water-containing sporocysts, whereas the definitive host becomes infected through predating or scavenging tissues containing mature sarcocysts [1,2].

Wild and domestic birds are intermediate hosts for 28 *Sarcocystis* species [3,4]. Some of these species are pathogenic, causing severe myositis, encephalitis, hepatitis, and pneumonia [5–10]. Birds of the order Anseriformes are known to be intermediate hosts of five *Sarcocystis* species: *Sarcocystis albifronsi*, *Sarcocystis anasi*, *Sarcocystis platyrhynchosi*, *Sarcocystis rileyi*, and *Sarcocystis wobeseri* [4,11–14]. Four of these species, *S. albifronsi*, *S. anasi*, *S. rileyi*, and *S. wobeseri*, have been detected in ducks and geese from Lithuania [12–15]. Only macrocysts of *S. rileyi* are visible to the naked eye [11,14–16], while other *Sarcocystis* species form microcysts in the muscles of anseriforms [4,12,13]. Although *S. rileyi* is not considered to be pathogenic, hunted birds infected with macrocysts of this species are not suitable for human consumption [17–22]. Therefore, extensive efforts have been made to investigate the epidemiology and determine hosts of *S. rileyi*.

For a long time, reports of macrocysts in the muscles of anseriforms came almost exclusively from North America. Macrocysts were observed in numerous duck species of the genera *Aix*, *Anas*, *Aythya*, *Bucephala*, *Mareca*, *Melanitta*, and *Spatula* [11,16,18,23–34]. Based on the first summary list of *Sarcocystis* spp. in birds, presented by Erickson in 1940, macrocysts of *Sarcocystis* spp. were detected in the Mallard (*Anas platyrhynchos*), American Black Duck (*Anas rubripes*), Gadwall (*Mareca strepera*), Northern Pintail (*Anas acuta*), Blue-Winged Teal (*Spatula discors*), and Northern Shoveler (*Spatula clypeata*) in North America [24]. Later in North America, macrocysts were recorded in the Green-Winged Teal (*Anas carolinensis*), Wood Duck (*Aix sponsa*), Greater Scaup (*Aythya marila*), Bufflehead (*Bucephala albeola*), American Wigeon (*Mareca americana*), Redhead (*Aythya americana*), Canvasback (*Aythya valisineria*), Common Goldeneye (*Bucephala clangula*) [26], Mottled Duck (*Anas fulvigula*), Lesser Scaup (*Aythya affinis*) [27], and Velvet Scoter (*Melanitta fusca*) [30].

In 2003, a detailed morphological characterization of macrocysts found in the Shoveler Duck (*Anus cylpeata*) was provided using light microscopy and transmission electron microscopy, and the parasite detected was redescribed as *S. rileyi* [11]. The first molecular characterization of *S. rileyi* was carried out in 2010 on macrocysts isolated from the Mallard collected in the USA [16]. Based on DNA sequencing of three genetic loci, macrocysts found in Mallards were confirmed as *S. rileyi* in Europe in 2011 [14]. Notably, reports on macrocysts in waterfowl, mainly Mallards, have increased in the twenty-first century in Europe [14,15,17,19–22,35,36].

Based on the experimental infection, the striped skunk (*Mephitis mephitis*) (Mephistidae family) was confirmed to be a definitive host of *S. rileyi* in North America [37,38]. The striped skunk is unsuitable for the distribution of *S. rileyi* in Europe since this small predatory animal lives on this continent only in captivity [39]. The molecular analysis enabled the identification of *S. rileyi* sporocysts in representatives of the families Canidae and Mustelidae in Europe. In Germany, *S. rileyi* was confirmed in the red fox (*Vulpes vulpes*) and in the raccoon dog (*Nyctereutes procyonoides*) [40]. In addition to these two predatory species, *S. rileyi* was also identified in the small intestines of the members of the family Mustelidae, in the American mink (*Neogale vison*), the European polecat (*Mustela putorius*), and the European badger (*Meles meles*) from Lithuania [41,42].

The prevalence of macrocysts of *Sarcocystis* in wild ducks depends on various factors, such as bird species, age, sex, location, etc., and has been broadly studied in North America [27,28,30]. However, detailed studies on *Sarcocystis* species identification, variety of intermediate hosts, life cycle, and infection rates of macrocysts in anseriforms in Europe are still lacking. Results of the first long-term (over a decade) surveillance study on macrocysts of *Sarcocystis* in wildfowl species in Europe (with more than 3000 anseriforms checked for macrocysts in Lithuania, Latvia, Russia, and Belarus) and molecular identification of *S. rileyi* in Mallards from four countries are presented in the present paper.

2. Materials and Methods

2.1. Sample Collection

For the evaluation of macrocysts in wildfowl, 2649 birds were collected in Lithuania in 2011–2014 and 2017–2022. Mallards accounted for the largest number of birds surveyed (2362), although 287 wildfowl belonging to other 12 species were also analyzed. The examined birds were collected from all counties in Lithuania.

The majority of the analyzed birds were legally hunted during the open hunting season (August–December) by licensed third parties, following the Hunting Law of the Republic of Lithuania. It should be noted that the birds were not hunted for the survey. Specifically, all examined Mallards, as well as 88 Eurasian Teals (*Anas crecca*), 46 Greater White-Fronted Geese (*Anser albifrons*), 29 Tufted Ducks (*Aythya fuligula*), 25 Garganeys (*Spatula querquedula*), 16 Common Goldeneyes, five Barnacle Geese (*Branta leucopsis*), and a single Bean Goose (*Anser fabalis*) were hunted. Furthermore, birds of several other wildfowl species found dead were received from Kaunas Tadas Ivanauskas Zoology Museum (the national authority responsible for monitoring wild birds found dead) and examined. This sample included 42 Velvet Scoters, 14 Gadwalls, 10 Long-Tailed Ducks (*Clangula hyemalis*), 10 Eurasian Wigeons (*Mareca penelope*), and a single Common Pochard (*Aythya ferina*). Carcasses of birds collected by hunters and staff from the Kaunas Tadas Ivanauskas Zoology Museum were frozen and delivered to scientific laboratories for examination of sarcocysts.

In addition, a total of 619 Mallards were received from neighboring countries. Specifically, 319 Mallards were delivered from the Kaliningrad Region of Russia, 202 birds came from Belarus, and 98 were sent from Latvia. All analyzed Mallards were hunted in the autumn seasons of 2012–2014 in accordance with the national hunting regulations valid in the countries concerned and checked for macrocysts of *Sarcocystis*.

The skin of all collected birds was peeled from the head to the area of the knee tendons, and the muscles were examined grossly for macrocysts of *Sarcocystis* spp. visible to the naked eye.

2.2. Molecular Examination of Macrocysts

During the study, the investigators presumed that macrocysts observed in ducks belonged to *S. rileyi*. For the confirmation of the parasite species, only some of the macrocysts isolated from the infected birds were subjected to DNA analysis. *Sarcocystis* species have not been molecularly identified in all infected ducks, mainly due to the fact that the research was carried out by two independent research groups and also taking financial issues into account.

During the period of 2011–2014, birds examined were transported to the Laboratory of Molecular Ecology of the Nature Research Centre (Vilnius, Lithuania). At that time, macrocysts isolated from 32 infected Mallards were subjected to DNA analysis for confirmation of *Sarcocystis* species. In all cases, a single macrocyst was excised from an individual bird. In total, two of the infected Mallards came from Latvia, six from Belarus, eight from the Kaliningrad Region of Russia, and the remaining sixteen birds were hunted in Lithuania.

Meanwhile, from 2017 to 2022, monitoring of macroscopic sarcocysts found in ducks hunted in Lithuania was carried out. The collected individuals of wildfowl species were submitted to the Laboratory of Forest Sciences Department of the Faculty of Forestry and Ecology of the Agriculture Academy of Vytautas Magnus University (Kaunas, Lithuania). The carcasses of birds were surveyed for the macrocysts of *Sarcocystis*, but only five macrocysts isolated from individual Mallards hunted in late autumn 2022 in Lithuania were examined by molecular methods. Thus, in total, macrocysts isolated from 37 Mallards were used for molecular identification of *Sarcocystis* species.

Prior to the molecular analysis, the macrocysts were stored in individual tubes filled with 75% ethyl alcohol at −20 °C. The genomic DNA from the sarcocysts was extracted using spin-column based commercial kits, "NucleoSpin Tissue, Mini kit" (Macherey-Nagel, Düren, Germany), "QIAamp® DNA Micro Kit" (Qiagen, Hilden, Germany), or "GeneJET Genomic DNA Purification Kit" (Thermo Fisher Scientific Baltics, Vilnius, Lithuania).

Regardless of the kit, 100 μL of commercial Elution Buffer was added in the last step of DNA purification to elute the genomic DNA. For *Sarcocystis* species identification, all the DNA samples were subjected to the amplification of internal transcribed spacer 1 (*ITS1*), situated between *18S* rRNA and *5.8S* rRNA. This molecular locus is considered to be the most variable for *Sarcocystis* spp. in birds [43]. Furthermore, eight DNA isolates obtained from macrocysts of Mallards hunted in Lithuania were genetically characterized at *18S* rDNA and *28S* rDNA, and four of these isolates were also characterized at mitochondrial cytochrome c oxidase I (*cox1*) and the RNA polymerase B gene of the apicoplast genome (*rpoB*). The oligonucleotide primers used for the amplification, conditions of PCRs, purification of amplified PCR products, and procedures of direct Sanger sequencing were performed as described previously [44]. The resulting sequences were manually checked for ambiguously placed nucleotides and merged into a single fragment in the case of *18S* rDNA. Finally, the obtained DNA fragments were checked for similarity using the nucleotide BLAST program [45].

2.3. Statistical Data Analysis

A 95% confidence interval (CI) for the determined prevalence of macrocysts in different years, host species, countries, sexes, and months was computed using Sterne's exact method [46]. A pair-wise comparison of infection rates of macrocysts was made by the unconditional exact test [47]. The following analysis is more sensitive in detecting differences than Fisher's exact test and is especially advantageous in the case of small samples ($n1$, $n2 < 100$). Fisher's exact test was used to compare the prevalence data of macrocysts between three and six samples, and the Chi-squared test was used when more than six samples were compared. Statistical analyses were performed in Quantitative Parasitology 3.0 software [48].

3. Results

3.1. The Occurrence of Macrocysts in Wildfowl Species and Parasite Load

Macroscopic sarcocysts were found almost exclusively in the pectoral muscles of birds, although they were also detected in the neck muscles of several individuals. The exact number of birds with macrocysts in their necks was not recorded.

Macrocysts of *Sarcocystis* were yellowish-white in color and 2.0–7.0 × 1.5–2.0 mm (1.8 × 4.6 mm) in size. They resembled rice grains in shape and color. It is noteworthy that the parasite load varied greatly in infected birds. Among all the examined individuals, several were heavily infected, some birds acquired moderate infection, and in the carcasses of other birds, only single macrocysts were detected (Figure 1). However, during this study, the parasite load in the muscles of birds was not examined in detail, and parasite load categorization according to the number of macrocysts observed was subjective.

3.2. The Prevalence of Macrocysts of Sarcocystis in Wildfowl in Lithuania and Neighboring Countries

Overall, macrocysts were found in 206 of 2362 (8.7%, CI = 7.65–9.93%) Mallards collected in Lithuania during different years (2011–2014 and 2017–2022). A similar prevalence was observed in the examined Mallards collected in Lithuania, the Kaliningrad Region of the Russian Federation, Belarus, and Latvia, from 3.1% (CI = 0.84–8.53%) in Latvia to 5.9% (CI = 3.37–10.08%) in Belarus (Table 1), although differences were not significant. However, significant differences were noticed when comparing the prevalence of infection in 2012 ($p = 0.005$) and 2014 ($p = 0.002$). In 2012, infection rates were much higher in the Kaliningrad Region of Russia ($p = 0.002$) and Belarus ($p = 0.010$) than those in Lithuania. Whereas in 2014, a significantly higher prevalence of macrocysts was recorded in Lithuania than that observed in the Kaliningrad Region of Russia ($p = 0.005$) and Belarus ($p = 0.008$) (Table 1).

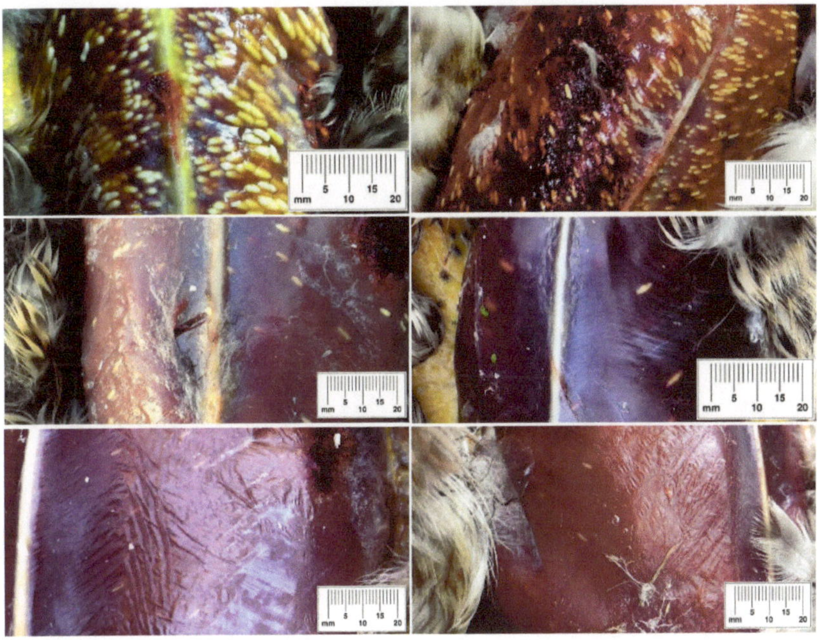

Figure 1. Different parasite loads of macrocysts of *Sarcocystis* in Mallard carcasses. Images are ordered by the size of the parasite load, from highest to lowest. The scale shows the size of macrocysts in mm.

Table 1. The prevalence of macrocysts in examined Mallards collected in four countries in 2012–2014.

Year	Lithuania	Kaliningrad Region (Russia)	Belarus	Latvia
2012	2/158 (1.3%) ^{b,e}	10/107 (9.4%) ^{c,f}	6/72 (8.3%) ^{g}	3/98 (3.1%)
2013	1/21 (4.8%)	2/74 (2.7%)	5/62 (8.1%)	
2014	9/66 (13.6%) ^{a,h}	4/138 (2.9%) ^{d,i}	1/68 (1.5%) ^{j}	
Overall	12/245 (4.9%)	16/319 (5.0%)	12/202 (5.9%)	3/98 (3.1%)

The statistically significant p values ($p < 0.05$) obtained by comparing the prevalence of macrocysts in four countries in different years are given below. $^{a} > ^{b}$ $p < 0.001$; $^{c} > ^{d}$ $p = 0.034$; $^{f} > ^{e}$ $p = 0.002$; $^{g} > ^{e}$ $p = 0.010$; $^{h} > ^{i}$ $p = 0.005$; $^{h} > ^{j}$ $p = 0.008$.

A significant difference ($\chi^2 = 17.97$, df = 9, $p = 0.035$) was observed when comparing the prevalence of macrocysts in Mallards collected in different years (2011–2014 and 2017–2022) in Lithuania (Figure 2). The difference remained significant ($\chi^2 = 15.70$, df = 7, $p = 0.028$) when samples containing fewer than 150 birds (2013 and 2014 years) were excluded. The lowest prevalence was estimated in 2012 (1.3%, CI = 0.23–4.60%), while the highest was in 2014 (13.6%, CI = 7.07–24.01%) (Figure 2). Prevalence was relatively consistent, varying from 7.7 to 11.9% in 2017–2022. The prevalence of macrocysts in 2012 was significantly lower than that determined in 2011 ($p = 0.002$), 2014 ($p < 0.001$), 2017 ($p < 0.001$), 2018 ($p < 0.001$), 2019 ($p < 0.001$), 2020 ($p = 0.001$), 2021 ($p < 0.001$), and 2022 ($p = 0.002$). Also, a low prevalence was determined in Mallards collected in 2013 (4.8%, CI = 0.2–23.3%); however, this data should be interpreted with caution due to the small sample size of the birds examined ($n = 21$).

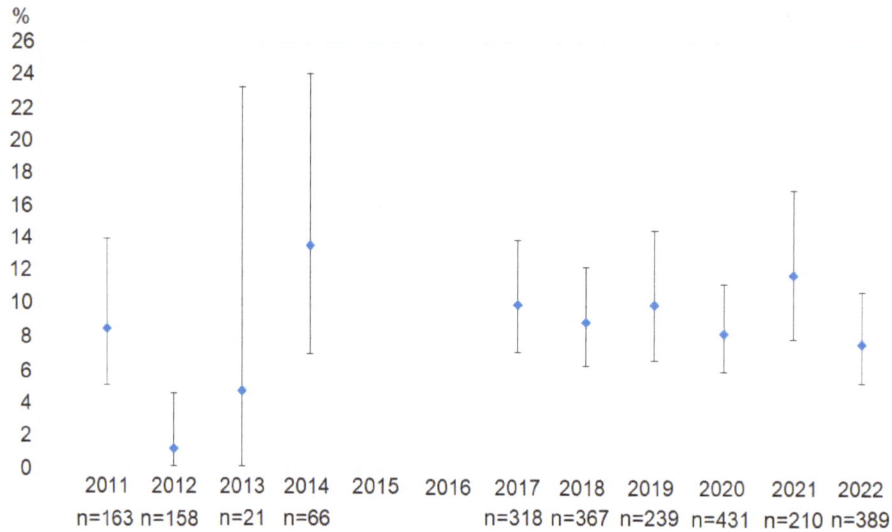

Figure 2. Prevalence of macrocysts in Mallards collected in Lithuania in August–December 2011–2014 and 2017–2022. Error bars indicate 95% confidence intervals. *n* shows the size of the sample in each analyzed year.

Out of the other 12 wildfowl species tested (287 individuals), macrocysts were detected only in the breast muscles of the Eurasian Teal, with a low prevalence (2.3%, 2 infected of 88 analyzed, CI = 0.41–7.79%). In Lithuania, a significantly higher ($p = 0.030$) prevalence of macrocysts was established in Mallards (206/2362, 8.7%) than that observed in Eurasian Teals (2/88, 2.3%).

3.3. *Factors Influencing the Prevalence of Macrocysts of Sarcocystis in the Mallard*

In 2011–2014 and 2017–2022, the lowest prevalence of macrocysts of *Sarcocystis* spp. in Mallards in Lithuania was noticed in October (7.4%, CI = 5.56–9.72%) and in September (7.7%, CI = 5.78–10.01%) (Figure 3), while the highest was observed in December (14.1%, CI = 8.07–23.40%). The infection rate in August (9.6%, CI = 7.63–11.83%) and in November (10.6%, CI = 7.05–15.31%) was similar. The prevalence was independent of the month when Mallards were hunted. However, a pair-wise analysis demonstrated a significantly higher prevalence in November–December than in September–October ($p = 0.028$). Also, a significantly higher prevalence of macrocysts was determined in November–December than in October ($p = 0.036$) and in December than in September–October ($p = 0.038$).

In Lithuania, during the hunting season (2011–2014 and 2017–2022), a higher but statistically insignificant prevalence was detected in Mallard males (99 infected of 1085 analyzed, 9.1%, CI = 7.55–10.99) than in females (107 infected of 1277 examined, 8.4%, CI = 6.98–10.03).

3.4. *Molecular Identification of S. rileyi in Mallards from Lithuania, Russia, Belarus, and Latvia*

Based on the comparison of the current work-generated DNA sequences, *S. rileyi* was confirmed in all tested Mallards, 21 from Lithuania, eight from the Kaliningrad Region of Russia, six from Belarus, and two from Latvia. Comparing the present work determined sequences of the five genetic loci, intraspecific variation was not observed.

Thirty-seven 942 bp-long sequences of the complete *ITS1* region were 100% identical to each other and showed 100% identity to *S. rileyi* previously identified in Mallard from Lithuania (HM185744) and to other European isolates of *S. rileyi* from ducks (KJ396584, MZ151434, MZ468639-40, LT992314). Single *ITS1* region sequences from four different countries have been submitted to GenBank under accession numbers OR416415–OR416418. The obtained *ITS1* sequences also displayed 99.9% identity to *S. rileyi* from Mallard collected in the USA (GU188427) and 99.5–100% identity to *S. rileyi* observed in the small intestines of the American mink, European polecat, and European badger (OP970971–81) from Lithuania.

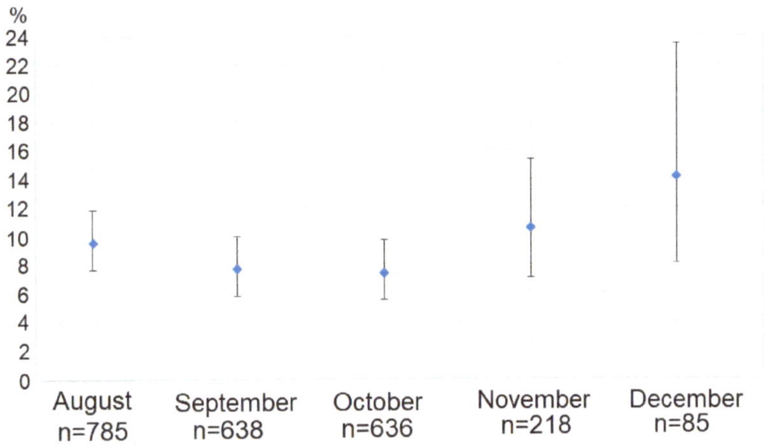

Figure 3. Monthly prevalence of macrocysts infection in Mallards in Lithuania from August to December. Dots represent the percentage infection rates. Error bars indicate 95% confidence intervals of prevalence. *n* indicates the sample size in the months studied.

In addition to the *ITS1* region, eight partial 1753 bp-long *18S* rDNA, eight 1499 bp-long *28S* rDNA, four 1053 bp-long *cox1*, and four 762 bp-long *rpoB* sequences were identified. As the sequences identified during this study were identical in the same genetic region, one each of *18S* rDNA (OR416136), *28S* rDNA (OR416169), *cox1* (OR423018), and *rpoB* (OR423019) were uploaded to the GenBank. In the current work, the obtained rDNA sequences were 100% identical with homological sequences of *S. rileyi* previously defined in Mallard from Lithuania (HM185742–43). Our 1753 bp-long *18S* rDNA sequences showed 100% identity with *S. rileyi* isolated from various countries and hosts (GU120092, KJ396583, KM233682, MZ151434, MZ468637–38), 99.8% identity with *S. rileyi* isolates obtained from Mallard, Gadwall European and Eurasian Wigeons in the UK (LT992317–23), and 99.8% identity with the 1010 bp-long sequence of *S. rileyi* (KX170885) obtained from Northern Shoveler in Mexico. Notably, sequence mismatches were observed only at the ends of the aligned *18S* rDNA sequences, suggesting that these differences are due to sequencing errors rather than representing intraspecific genetic variability. The 1499 bp-long *28S* rDNA sequences determined in the present work demonstrated 100% identity to *S. rileyi* from Mallards in Denmark (MZ151434, MZ468640–41) from Mallard in the USA (GU188426) and 99.9% identity to *S. rileyi* from Common Eider in Norway (KJ396585). The 762 bp-long *rpoB* sequence determined in the present study showed 100% identity with that of *S. rileyi*, previously also isolated from Mallard in Lithuania (MF596308). Finally, our 1053 bp-long *cox1* sequence displayed 100% similarity compared to *S. rileyi* from Common Eider in Norway (KJ396582).

In terms of interspecific genetic variability, the largest differences between *S. rileyi* and the most related species, *Sarcocystis atraii*, *Sarcocystis wenzeli*, *Sarcocystis cristata*, *Sarcocystis chloropusae*, *S. anasi* and *S. albifronsi* [3,49,50], were observed in the *ITS1* region (9.9–25.7%). Comparing *S. rileyi* with the listed *Sarcocystis* species, much less variability was determined within *cox1* (2.4–3.6%), *rpoB* (2.6–4.1%) and *28S* rDNA (0.6–3.8%) while differences in *18S* rDNA were minimal (0.3–1.1%).

4. Discussion

4.1. The Genetic Identification and Variability of S. rileyi

Among hunters and veterinarians, *S. rileyi* is known as a parasite that forms "rice-like grains". Macrocysts were discovered in a duck by Riley and were named *S. rileyi* by Stiles in 1893 [51]. Thus, *S. rileyi* is the first *Sarcocystis* species discovered in birds and, in general, one of the first *Sarcocystis* species described [1]. Researchers finding macrocysts in the muscles of waterfowl attributed them to *S. rileyi* [11,23,24,26]. Taking into account that five *Sarcocystis* spp. are known to form sarcocysts in the muscles of anseriforms [4], the possibility that more than one species of *Sarcocystis* may constitute macrocysts in the muscles of ducks cannot be excluded [16]. Therefore, molecular methods are needed for the identification of *S. rileyi* [11,16]. In the current investigation, macrocysts were found in 239 individuals of two wildfowl species (237 Mallards and two Eurasian Teals). Due to the limitation of this study that molecular methods were only used on the part of the infected birds, we cannot claim that all infected ducks harbored macrocysts of *S. rileyi*. Overall, thirty-seven macrocysts were molecularly analyzed, and *S. rileyi* was identified in twenty-one Mallards from Lithuania, eight Mallards from Russia's Kaliningrad Region, six Mallards from Belarus, and two Mallards from Latvia. To the best of our knowledge, this is the first report of *S. rileyi* in Russia, Belarus, and Latvia. Notably, macrocysts have previously been found in the muscles of ducks from Russia, but no molecular investigations have been carried out on the parasites detected [52,53]. Based on the recent reports on *S. rileyi* in various European countries [14,15,17,19–22,35,36] and records of experienced local hunters, macrocysts of *Sarcocystis* are assumed to be spreading geographically [22]. In Lithuania, *S. rileyi* in Mallards has been previously confirmed using DNA sequence analysis [14,15]. Summarizing the results of this research and previous studies carried out in Lithuania, all 40 genetically examined macrocysts isolated from different Mallard individuals were identified as *S. rileyi* [14,15; and present work]. In light of these data, we believe that most of the macrocysts found in the muscles of Mallards hunted in Lithuania should belong to *S. rileyi*.

Based on the results of the present and earlier studies [14–16,21,22,43,54], the *ITS1* region is the most suitable loci for the identification of *S. rileyi*, as significantly less inter-specific variation is observed when comparing nuclear sequences of the *18S* rDNA, *28S* rDNA, mitochondrial *cox1*, and apicoplast *rpoB* loci. Recently, the complete *ITS2* region of *S. rileyi*, which is located between *5.8S* rDNA and *28S* rDNA, has been reported [22]. This genetic marker has good potential for the identification of species in the Sarcocystidae family [55]. In the current study, no intraspecific variation has been observed when analyzing sequenced isolates of *S. rileyi*. The restricted genetic diversity of *S. rileyi* has been reported previously [15]. Among the genetic regions currently used most frequently for the characterization of *Sarcocystis* species employing birds as intermediate hosts [3,56], *ITS1* is known to be the most variable at the intraspecific level [57]. The current data available in GenBank show low intraspecific variability of *S. rileyi* within *ITS1*. Intriguingly, high intraspecific genetic variation was observed for *Sarcocystis halieti* [54,58] and *Sarcocystis falcatula* [59–61] using birds of several different orders as their intermediate hosts. Thus, the relatively low intraspecific genetic variability of *S. rileyi* might be explained by the high intermediate host specificity, as macrocysts of *S. rileyi* have only been validated in ducks of the family Anatidae [11,14–22].

4.2. The Distribution of Sarcocystis Macrocysts in Different Wildfowl Species

Summarizing record data on macrocysts of *Sarcocystis* spp. in wildfowl species in North America, this infection occurs significantly more often in dabbling ducks than in diving ones [27,32]. Dabbling ducks feed in shallower wetlands and upland regions, so they have a greater chance of ingesting the fecal material of predators than divers do [28]. Apart from North America, macrocysts have been found in ducks in Europe [14,15,17,19–22,35,36,62] but not in Australia or Asia [63,64], while ducks were not examined for *Sarcocystis* spp. in detail in South America and Africa [1,11,16].

The range of intermediate hosts of macrocysts of *Sarcocystis* is considerably smaller in Europe than in North America. In Europe, macrocysts were mostly recorded in Mallards [14,17,19–22,35,62], Common Eider (*Somateria mollissima*) in Norway [36], Eurasian Wigeon, Eurasian Teal in Finland and the UK [15,21], in the Gadwall, and Northern Pintail in the UK [21]. In the present study, macrocysts were observed in Mallards and Eurasian Teals collected in Lithuania. This is the first survey in which macrocysts of *Sarcocystis* have been detected in European Teal Lithuania. In the previous study conducted in Lithuania, macrocysts were found only in Mallards [14,15,62]. In the present study, macrocysts were not noticed in other wildfowl species analyzed. The non-detection of macrocysts in other bird species can be explained by the small number of individuals examined (n = 1–46). However, taking into account records of macrocysts and data on infection rates in this study and previous investigations [14,15,21], it is suggested that the Mallard is the main source of macrocyst infection in Europe. Nevertheless, it cannot be ruled out that the highest number of detections of macrocysts in Mallards in Europe [15,21] is related to the fact that this species is the most hunted of all the anseriforms [65,66]. On the contrary, in North America, the prevalence of macrocysts in Mallards was not the highest as compared to that in Northern Pintails, Blue-winged Teals, Green-Winged Teals, American Wigeons, Northern Shovelers [27,28], Lesser Scaups, Velvet Scoters [30], American Black Ducks [32], and Mottled Ducks [34]. Thus, in North America, compared to Europe, macrocysts are more common in various duck species [11,16,18,23–34]. A smaller range of intermediate hosts in Europe, compared to North America, may indicate that in Europe, the infection of macrocysts of *Sarcocystis* spp. has spread recently [14,15,17,19–22,35,36,62].

4.3. Factors Affecting the Prevalence of Macrocysts of *Sarcocystis* spp. in the Mallard in Lithuania

The first report on macrocysts in Mallards in Europe was recorded in Russia in the middle of the 20th century [52,53]. Reports of macrocysts of *Sarcocystis* in Europe have increased in the 21st century. Macrocysts in ducks were detected in Lithuania [14,15,62], Poland [17], the Slovak Republic [35], Finland [15], Norway [36], the Czech Republic, Sweden, the Netherlands, Austria [for details see 15], Hungary [20], UK [19,21], and Denmark [22]. Thus, macrocysts were most commonly reported in northern, eastern, and central Europe. Based on molecular results, the invasive American mink is one of the definitive hosts of *S. rileyi* [42]. This predator of the family Mustelidae is native to North America, and its introduced range [67,68] partially coincides with the geography of the distribution of macrocysts of *Sarcocystis* spp. in Europe. Therefore, it can be speculated that the spread of *S. rileyi* in Europe is associated with the introduction of American mink to this continent.

In the present study, the exceptionally low prevalence of macrocysts (1.3%, CI = 0.23–4.60%) was determined in Mallards hunted in 2012 in Lithuania (Figure 3). Infection rates determined in 2012 were significantly lower than those observed in 2011, 2014, and 2017–2022. Furthermore, infection rates of macrocysts of *Sarcocystis* spp. observed in Lithuania and Kaliningrad/Belarus differed significantly in 2012 and 2014 (Table 1). These findings can be partly accounted for by the different periods when Mallards were hunted in the countries concerned. In Lithuania, the majority of the ducks analyzed in 2012 were hunted in August–early September (with local breeders forming the majority of the analyzed ducks), whereas in Belarus and the Kaliningrad Region of Russia, ducks were hunted in August–December 2012. Some Mallards staging and wintering in Lithuania, Belarus, and the Kaliningrad Region of Russia in October–December are breeders from Northern Europe [69]. A higher prevalence of macrocysts detected in Belarus and the Kaliningrad Region of Russia in 2012 is likely to have been caused by a larger proportion of the analyzed migrants from the major region of Northern Europe [70,71]. It should be noted that a different prevalence of macrocysts was also observed in different geographical areas of the USA and Canada [27,30]. The data obtained was explained by the different migratory routes of the infected birds and the abundance of predators in the breeding and wintering grounds of ducks [30].

To the best of our knowledge, the current study is the first to evaluate the prevalence of macrocysts of *Sarcocystis* spp. in intermediate hosts in different months. We demonstrated that the infection rate of macrocysts in Mallards hunted in Lithuania was significantly higher in November–December than in Mallards hunted in September–October. The data obtained can be explained by the fact that in November–December, migrants from Northern Europe form a significant proportion of all Mallards recorded in Lithuania [65,66]. However, to test the above assumption, further research on this infection is needed by comparing epidemiological data on parasites with genetic data on Mallard populations.

In this study, we found no significant differences when comparing the detection of macrocysts in the muscles of female and male Mallards. Thus, our findings are in agreement with previous studies showing that the prevalence of macrocysts does not depend on the sex of the birds examined [27,30].

4.4. The Significance of Macrocyst Infection in Ducks

Previous studies have demonstrated that some ducks infected with macrocysts of *Sarcocystis* show signs of myopathy [5–10]; however, histological examination does not show that the muscles of ducks are severely affected [21]. Even though this infection is not considered pathogenic, it may interfere with the energy and protein metabolism of birds [33]. Furthermore, severe infections of macrocysts in muscles used for locomotion may cause a reduced flying capacity and ability to withstand migration, weakness, and an increased probability of becoming a victim of predators [72].

Significant economic losses are incurred worldwide as a result of macrocysts of *Sarcocystis* spp. [21]. The Mallard is the world's most abundant and widely distributed dabbling duck species, whose global population is estimated at 20 million individuals [66]. In Lithuania, it is the most abundant breeding wildfowl species, with an estimated population of about 60,000 pairs [71]. The Mallard is one of the most harvested game species worldwide, with more than 4.5 million birds annually hunted in Europe [65]. Activities related to the hunting of the Mallard constitute an important part of income in the recreation, tourism, agriculture, food, and catering sectors. Meat from hunted ducks infected with macrocysts of *Sarcocystis* is not suitable for human consumption [1,15,19,21,22,35,36]. The toxic effects of the parasites can be harmful when eating heavily infected ducks [30], though properly cooked meat from infected ducks is considered to be harmless to humans [73]. Therefore, awareness related to the infection of macrocysts must be increased among hunters, veterinarians, and the general public. Infection by *Sarcocystis* spp. parasites and their prevention are mentioned among the biological contamination risk factors specified in the game meat safety requirements valid in Lithuania and in other countries of the European Union. There is a lack of such strict regulations in the countries of Eastern Europe, where large quantities of meat from hunted ducks are consumed in the households of hunters. Furthermore, changes in game eating habits are observed; dishes are increasingly being prepared from raw, uncooked game meat. Therefore, hunters should be able to recognize macroscopic sarcocysts and know how to utilize the infected birds [19,21,22].

The Mallard can also form a potential link between domestic birds and wild ducks in spreading parasites of the genus *Sarcocystis*. During migration periods, staging Mallards are recorded in numerous pools and ponds used for the production of domestic ducks and geese in various regions of Eastern Europe [71]. These ducks are very tolerant of human presence and are common in urban and agricultural habitats, thus forming a potential link between wild waterfowl, poultry, and humans [74]. More than three million Mallards reared in captivity are released annually for hunting purposes in Europe [75]. There is a lack of knowledge about the infection of macrocysts of *Sarcocystis* in poultry, as reports on macrocysts in domestic ducks are very scarce [24,53]. Therefore, it is important to undertake studies into the potential infection of macrocysts in poultry, particularly among the millions of farmed Mallards that are released for hunting purposes in Europe each year.

5. Conclusions

Sarcocystis rileyi was, for the first time, identified in Mallards hunted in the Kaliningrad region of the Russian Federation, Belarus, and Latvia based on DNA sequencing analysis of some of the birds infected with macrocysts. Findings of the present work show that infection rates of macrocysts of *Sarcocystis* in the muscles of ducks depend on the host species, hunting year, geographical region, and hunting season.

The Mallard is the main source of infection by macrocysts in Europe. Since the Mallard is one of the most harvested game bird species worldwide and its meat contaminated with macrocysts of *Sarcocystis* is not suitable for consumption, awareness of the infection of macrocysts must be increased. Hunted ducks should be tested for rice-like macrocysts, and it is recommended that hunted macrocyst-infected ducks not be left in the wild, consumed as food, or fed to domestic predators. Also, further studies on the macrocysts of *Sarcocystis* spp. in farmed ducks are needed.

Author Contributions: Conceptualization, P.P., J.S., D.B. and R.V.-B.; methodology, P.P., J.S. and R.V.-B.; software, P.P. and J.S.; validation, J.S. and S.Š.; formal analysis, P.P., J.S. and S.Š.; investigation, P.P., D.B., J.S., R.V.-B. and E.J.-N.; resources, D.B. and R.V.-B.; data curation, P.P. and J.S.; writing—original draft preparation, P.P., J.S. and S.Š.; writing—review and editing, P.P., J.S., S.Š., D.B., R.V.-B. and E.J.-N.; visualization, P.P., J.S. and R.V.-B.; supervision, P.P. and R.V.-B. All authors have read and agreed to the published version of the manuscript.

Funding: Part of the study conducted in the years 2013–2015 was conducted within the project "Investigations into genus *Sarcocystis* parasites from birds order Anseriformes", funded by the Research Council of Lithuania (Grant No. MIP-55/2013) and headed by Dr. Petras Prakas.

Institutional Review Board Statement: Samples were collected with the permission of the Ministry of Environment of the Republic of Lithuania (19 March 2013 no. 17; 31 March 2014 no. 15; 23 March 2017 no. 26-A4-3119; 1 March 2019 no. 26-A4-1535; 31 March 2021 no. 26-SR-89).

Informed Consent Statement: Not applicable.

Data Availability Statement: The sequences of *S. rileyi* generated in the current research were submitted to the NCBI GenBank database under accession numbers OR416136, OR416169, OR416415–OR416418, OR423018, OR423019.

Acknowledgments: The authors are grateful to Alexander Kozulin, Gennady Grishanov, and Mara Janaus for providing bird samples for this study collected in Belarus, the Kaliningrad Region of Russia, and Latvia. We would like to express our sincere gratitude to numerous Lithuanian hunters who have regularly cooperated with scientists in monitoring *S. rileyi* in wildfowls.

Conflicts of Interest: The authors declare no conflict of interest.

References

1. Dubey, J.P.; Calero-Bernal, R.; Rosenthal, B.M.; Speer, C.A.; Fayer, R. *Sarcocystosis of Animals and Humans*, 2nd ed.; CRC Press: Boca Raton, FL, USA, 2016.
2. Fayer, R. *Sarcocystis* spp. in Human Infections. *Clin. Microbiol. Rev.* **2004**, *17*, 894–902. [CrossRef]
3. Máca, O.; González-Solís, D. *Sarcocystis cristata* sp. nov. (Apicomplexa, Sarcocystidae) in the Imported Great Blue Turaco *Corythaeola cristata* (Aves, Musophagidae). *Parasit. Vectors* **2021**, *14*, 56. [CrossRef]
4. Hu, J.; Zhang, M.; Wu, Z.; Zeng, H.; Tao, J. Description of *Sarcocystis platyrhynchosi* n. sp. (Apicomplexa: Sarcocystidae) from Domestic Ducks *Anas platyrhynchos* (Anseriformes: Anatidae) in China. *Parasit. Vectors* **2023**, *16*, 50. [CrossRef]
5. Smith, J.H.; Neill, P.J.; Box, E.D. Pathogenesis of *Sarcocystis falcatula* (Apicomplexa: Sarcocystidae) in the Budgerigar (*Melopsittacus undulatus*). III. Pathologic and Quantitative Parasitologic Analysis of Extrapulmonary Disease. *J. Parasitol.* **1989**, *75*, 270–287. [CrossRef]
6. Olias, P.; Gruber, A.D.; Heydorn, A.O.; Kohls, A.; Mehlhorn, H.; Hafez, H.M.; Lierz, M. A Novel *Sarcocystis*-Associated Encephalitis and Myositis in Racing Pigeons. *Avian Pathol.* **2009**, *38*, 121–2128. [CrossRef] [PubMed]
7. Wünschmann, A.; Rejmanek, D.; Cruz-Martinez, L.; Barr, B.C. *Sarcocystis falcatula*-Associated Encephalitis in a Bree-Ranging Great Horned Owl (*Bubo virginianus*). *J. Vet. Diagn. Investig.* **2009**, *21*, 283–287. [CrossRef] [PubMed]
8. Olias, P.; Gruber, A.D.; Heydorn, A.O.; Kohls, A.; Hafez, H.M.; Lierz, M. Unusual Biphasic Disease in Domestic Pigeons (*Columba livia* f. *domestica*) Following Experimental Infection with *Sarcocystis calchasi*. *Avian Dis.* **2010**, *54*, 1032–1037. [CrossRef]

9. Olias, P.; Gruber, A.D.; Kohls, A.; Hafez, H.M.; Heydorn, A.O.; Mehlhorn, H.; Lierz, M. *Sarcocystis* Species Lethal for Domestic Pigeons. *Emerg. Infect. Dis.* **2010**, *16*, 497–499. [CrossRef]
10. Wünschmann, A.; Rejmanek, D.; Conrad, P.A.; Hall, N.; Cruz-Martinez, L.; Vaughn, S.B.; Barr, B.C. Natural Fatal *Sarcocystis falcatula* Infections in Free-Ranging Eagles in North America. *J. Vet. Diagn. Investig.* **2010**, *22*, 282–289. [CrossRef] [PubMed]
11. Dubey, J.P.; Cawthorn, R.J.; Speer, C.A.; Wobeser, G.A. Redescription of the Sarcocysts of *Sarcocystis rileyi* (Apicomplexa: Sarcocystidae). *J. Eukaryot. Microbiol.* **2003**, *50*, 476–482. [CrossRef]
12. Kutkienė, L.; Prakas, P.; Sruoga, A.; Butkauskas, D. Description of *Sarcocystis anasi* sp. nov. and *Sarcocystis albifronsi* sp. nov. in Birds of the Order Anseriformes. *Parasitol. Res.* **2012**, *110*, 1043–1046. [CrossRef]
13. Kutkienė, L.; Prakas, P.; Sruoga, A.; Butkauskas, D. The mallard duck (*Anas platyrhynchos*) as intermediate host for *Sarcocystis wobeseri* sp. nov. from the barnacle goose (*Branta leucopsis*). *Parasitol. Res.* **2010**, *107*, 879–888. [CrossRef]
14. Kutkienė, L.; Prakas, P.; Sruoga, A.; Butkauskas, D. Identification of *Sarcocystis rileyi* from the Mallard Duck (*Anas platyrhynchos*) in Europe: Cyst Morphology and Results of DNA Analysis. *Parasitol. Res.* **2011**, *108*, 709–714. [CrossRef]
15. Prakas, P.; Oksanen, A.; Butkauskas, D.; Sruoga, A.; Kutkienė, L.; Švažas, S.; Isomursu, M.; Liaugaudaitė, S. Identification and Intraspecific Genetic Diversity of *Sarcocystis rileyi* from Ducks, *Anas* spp., in Lithuania and Finland. *J. Parasitol.* **2014**, *100*, 657–661. [CrossRef]
16. Dubey, J.P.; Rosenthal, B.M.; Felix, T.A. Morphologic and Molecular Characterization of the Sarcocysts of *Sarcocystis rileyi* (Apicomplexa: Sarcocystidae) from the Mallard duck (*Anas platyrhynchos*). *J. Parasitol.* **2010**, *96*, 765–770. [CrossRef]
17. Kalisinska, E.; Betlejewska, K.M.; Schmidt, M.; Gozdzicka-Jozefiak, A.; Tomczyk, G. Protozoal Macrocysts in the Skeletal Muscle of a Mallard duck in Poland: The First Recorded Case. *Acta Parasitol.* **2003**, *48*, 1–5.
18. Padilla-Aguilar, P.; Romero-Callejas, E.; Osorio-Sarabia, D.; Ramírez-Lezama, J.; Cigarroa-Toledo, N.; Machain-Williams, C.; Manterola, C.; Zarza, H. Detection and Molecular Identification of *Sarcocystis rileyi* (Apicomplexa: Sarcocystidae) From a Northern Shoveler (*Anas clypeata*) in Mexico. *J. Wildl. Dis.* **2016**, *52*, 931–935. [CrossRef] [PubMed]
19. Cromie, R.; Ellis, M. *Sarcocystis* Survey. *Sarcocystis* Survey Feedback Report—The UK Wildfowl *Sarcocystis* Survey. 2019. Available online: www.sarcocystissurvey.org.uk/2015-2018-feedback-report/ (accessed on 15 December 2022).
20. Szekeres, S.; Juhász, A.; Kondor, M.; Takács, N.; Sugár, L.; Hornok, S. *Sarcocystis rileyi* Emerging in Hungary: Is Rice Breast Disease Underreported in the Region? *Acta Vet. Hung.* **2019**, *67*, 401–406. [CrossRef] [PubMed]
21. Muir, A.; Ellis, M.; Blake, D.P.; Chantrey, J.; Strong, E.A.; Reeves, J.P.; Cromie, R.L. *Sarcocystis rileyi* in UK Free-living Wildfowl (Anatidae): Surveillance, Histopathology and First Molecular Characterisation. *Vet. Rec.* **2020**, *15*, 186. [CrossRef] [PubMed]
22. Zuo, S.; Sørensen, S.R.; Kania, P.W.; Buchmann, K. *Sarcocystis rileyi* (Apicomplexa) in *Anas platyrhynchos* in Europe with a Potential for Spread. *Int. J. Parasitol. Parasites Wildl.* **2021**, *15*, 270–275. [CrossRef] [PubMed]
23. Gower, C. A New Host and Locality Record for *Sarcocystis rileyi* (Stiles, 1803). *J. Parasitol.* **1938**, *24*, 378. [CrossRef]
24. Erickson, A.B. *Sarcocystis* in Birds. *Auk* **1940**, *57*, 514–519. [CrossRef]
25. Wichware, A.B. Sarcosporidiosis in a Black Duck. *Can. J. Comp. Med.* **1944**, *8*, 15–17.
26. Cornwell, G. New Waterfowl Host Records for *Sarcocystis rileyi* and a Review of Sarcosporidiosis in Birds. *Avian Dis.* **1963**, *7*, 212–216. [CrossRef] [PubMed]
27. Chabreck, R.H. Sarcosporidiosis in Ducks in Louisiana. *Trans. North Am. Wildl. Conf.* **1965**, *30*, 174–184.
28. Hoppe, D.M. Prevalence of Macroscopically Detectable *Sarcocystis* in North Dakota ducks. *J. Wildl. Dis.* **1976**, *12*, 27–29. [CrossRef]
29. Broderson, D.; Canaris, A.G.; Bristol, J.R. Parasites of Waterfowl from Southwest Texas: II. The Shoveler, *Anas clypeata*. *J. Wildl. Dis.* **1977**, *13*, 435–439. [CrossRef] [PubMed]
30. Drouin, T.E.; Mahrt, J.L. The Prevalence of *Sarcocystis* Lankester, 1882, in some Bird Species in Western Canada, with Notes on its Life Cycle. *Can. J. Zool.* **1979**, *57*, 1915–1921. [CrossRef] [PubMed]
31. Canaris, A.G.; Mena, A.C.; Bristol, J.R. Parasites of Waterfowl from Southwest Texas: III. The Green-Winged Teal, *Anas crecca*. *J. Wildl. Dis.* **1981**, *17*, 57–64. [CrossRef]
32. Costanzo, G.R. *Sarcocystis* in American Black Ducks Wintering in New Jersey. *J. Wildl. Dis.* **1990**, *26*, 387–389. [CrossRef] [PubMed]
33. Moorman, T.E.; Baldassarre, G.A.; Richard, D.M. The Frequency of *Sarcocystis* spp. and its Effect on Winter Carcass Composition of Mottled Ducks. *J. Wildl. Dis.* **1991**, *27*, 491–493. [CrossRef] [PubMed]
34. Fedynich, A.M.; Pence, D.B. *Sarcocystis* in Mallards on the Southern High Plains of Texas. *Avian Dis.* **1992**, *36*, 1067–1069. [CrossRef] [PubMed]
35. Hermanovská, L.; Čurlík, J.; Hurníková, Z.; Brunčák, J.; Oberhauserová, K.; Dvorožňáková, E. Sarcocystosis of Mallards in Slovak Respublic. *Folia Vet.* **2013**, *57*, 45–48.
36. Gjerde, B. Molecular Characterisation of *Sarcocystis rileyi* from a Common eider (*Somateria mollissima*) in Norway. *Parasitol. Res.* **2014**, *113*, 3501–3509. [CrossRef] [PubMed]
37. Cawthorn, R.J.; Rainnie, D.; Wobeser, G. Experimental Transmission of *Sarcocystis* sp. (Protozoa: Sarcocystidae) between the Shoveler (*Anas clypeata*) Duck and the Striped skunk (*Mephitis mephitis*). *J. Wildl. Dis.* **1981**, *17*, 389–394. [CrossRef]
38. Wicht, R.J. Transmission of *Sarcocystis rileyi* to the Striped skunk (*Mephitis mephitis*). *J. Wildl. Dis.* **1981**, *17*, 387–388. [CrossRef]
39. Wilson, D.E.; Reeder, D.M. *Mammal Species of the World: A Taxonomic and Geographic Reference*, 3rd ed.; Johns Hopkins University Press: Baltimore, MD, USA, 2005.
40. Moré, G.; Maksimov, A.; Conraths, F.J.; Schares, G. Molecular Identification of *Sarcocystis* spp. in Foxes (*Vulpes vulpes*) and Raccoon dogs (*Nyctereutes procyonoides*) from Germany. *Vet. Parasitol.* **2016**, *220*, 9–14. [CrossRef]

41. Prakas, P.; Liaugaudaitė, S.; Kutkienė, L.; Sruoga, A.; Švažas, S. Molecular Identification of *Sarcocystis rileyi* Sporocysts in Red foxes (*Vulpes vulpes*) and raccoon dog (*Nyctereutes procyonoides*) in Lithuania. *Parasitol. Res.* **2015**, *114*, 1671–1676. [CrossRef]
42. Prakas, P.; Moskaliova, D.; Šneideris, D.; Juozaitytė-Ngugu, E.; Maziliauskaitė, E.; Švažas, S.; Butkauskas, D. Molecular Identification of *Sarcocystis rileyi* and *Sarcocystis* sp. (Closely Related to *Sarcocystis wenzeli*) in Intestines of Mustelids from Lithuania. *Animals* **2023**, *13*, 467. [CrossRef]
43. Prakas, P.; Butkauskas, D.; Švažas, S.; Stanevičius, V. Morphological and Genetic Characterisation of *Sarcocystis halieti* from the Great Cormorant (*Phalacrocorax carbo*). *Parasitol. Res.* **2018**, *117*, 3663–3667. [CrossRef] [PubMed]
44. Kirillova, V.; Prakas, P.; Calero-Bernal, R.; Gavarāne, I.; Fernández-García, J.L.; Martínez-González, M.; Rudaitytė-Lukošienė, E.; Martínez-Estéllez, M.Á.H.; Butkauskas, D.; Kirjušina, M. Identification and Genetic Characterization of *Sarcocystis arctica* and *Sarcocystis lutrae* in Red Foxes (*Vulpes vulpes*) from Baltic States and Spain. *Parasit. Vectors* **2018**, *11*, 173. [CrossRef]
45. Altschul, S.F.; Madden, T.L.; Schäffer, A.A.; Zhang, J.; Zhang, Z.; Miller, W.; Lipman, D.J. Gapped BLAST and PSI-BLAST: A New Generation of Protein Database Search Programs. *Nucleic Acids Res.* **1997**, *17*, 3389–3402. [CrossRef] [PubMed]
46. Reiczigel, J. Confidence Intervals for the Binomial Parameter: Some New Considerations. *Stat. Med.* **2003**, *22*, 611–621. [CrossRef] [PubMed]
47. Reiczigel, J.; Abonyi-Tóth, Z.; Singer, J. An Exact Confidence Set for two Binomial Proportions and Exact Unconditional Confidence Intervals for the Difference and Ratio of Proportions. *Comput. Stat. Data Anal.* **2008**, *52*, 5046–5053. [CrossRef]
48. Rózsa, L.; Reiczigel, J.; Majoros, G. Quantifying Parasites in Samples of Hosts. *J. Parasitol.* **2000**, *86*, 228–232. [CrossRef] [PubMed]
49. El-Morsey, A.; El-Seify, M.; Desouky, A.R.; Abdel-Aziz, M.M.; El-Dakhly, K.M.; Kasem, S.; Abdo, W.; Haridy, M.; Sakai, H.; Yanai, T. Morphologic and Molecular Characteristics of *Sarcocystis atraii* n. sp. (Apicomplexa: Sarcocystidae) infecting the Common Coot (*Fulica atra*) from Egypt. *Acta Parasitol.* **2015**, *60*, 691–699. [CrossRef]
50. Pan, J.; Ma, C.; Huang, Z.; Ye, Y.; Zeng, H.; Deng, S.; Hu, J.; Tao, J. Morphological and Molecular Characterization of *Sarcocystis wenzeli* in Chickens (*Gallus gallus*) in China. *Parasit. Vectors* **2020**, *13*, 512. [CrossRef]
51. Stiles, C.W. On the presence of sarcosporidia in birds. *USDA Bureau Anim. Ind. Bull.* **1893**, *3*, 79–89.
52. Dylko, N.I. Distribution of Sarcosporidia among Wild Vertebrates in Belorussia. *Dokl. Akad. Nauk BSSR* **1962**, *6*, 399–400.
53. Golubkov, V.I. Infection of the Dog and Cat with Sarcocysts from the Chicken and Duck. *Veterinariya* **1979**, *1*, 55–56.
54. Prakas, P.; Butkauskas, D.; Juozaitytė-Ngugu, E. Molecular Identification of four *Sarcocystis* Species in the Herring Gull, *Larus argentatus*, from Lithuania. *Parasit. Vectors* **2020**, *13*, 2. [CrossRef]
55. Lee, F.C.H.; Muthu, V. From *18S* to *28S* rRNA Gene: An Improved Targeted Sarcocystidae PCR Amplification, Species Identification with Long DNA Sequences. *Am. J. Trop. Med. Hyg.* **2021**, *104*, 1388–1393. [CrossRef]
56. Prakas, P.; Butkauskas, D.; Juozaitytė-Ngugu, E. Molecular and Morphological Description of *Sarcocystis kutkienae* sp. nov. from the Common Raven (*Corvus corax*). *Parasitol. Res.* **2020**, *119*, 4205–4210. [CrossRef]
57. Šukytė, T.; Butkauskas, D.; Juozaitytė-Ngugu, E.; Švažas, S.; Prakas, P. Molecular Confirmation of *Accipiter* Birds of Prey as Definitive Hosts of Numerous *Sarcocystis* Species, including *Sarcocystis* sp., Closely Related to Pathogenic *S. calchasi*. *Pathogens* **2023**, *12*, 752. [CrossRef] [PubMed]
58. Juozaitytė-Ngugu, E.; Prakas, P. The Richness of *Sarcocystis* Species in the Common Gull (*Larus canus*) and Black-Headed Gull (*Larus ridibundus*) from Lithuania. *Parasitologia* **2023**, *3*, 172–180. [CrossRef]
59. Gondim, L.F.P.; Soares, R.M.; Tavares, A.S.; Borges-Silva, W.; de Jesus, R.F.; Llano, H.A.B.; Gondim, L.Q. *Sarcocystis falcatula*-like derived from Opossum in Northeastern Brazil: In Vitro Propagation in Avian Cells, Molecular Characterization and Bioassay in Birds. *Int. J. Parasitol. Parasites Wildl.* **2019**, *10*, 132–137. [CrossRef]
60. Llano, H.A.B.; Zavatieri Polato, H.; Borges Keid, L.; Ferreira de Souza Oliveira, T.M.; Zwarg, T.; de Oliveira, A.S.; Sanches, T.C.; Joppert, A.M.; Gondim, L.F.P.; Martins Soares, R. Molecular screening for Sarcocystidae in Muscles of Wild Birds from Brazil suggests a Plethora of Intermediate Hosts for *Sarcocystis falcatula*. *Int. J. Parasitol. Parasites Wildl.* **2022**, *17*, 230–238. [CrossRef]
61. Sato, A.P.; da Silva, T.C.E.; de Pontes, T.P.; Sanches, A.W.D.; Prakas, P.; Locatelli-Dittrich, R. Molecular Characterization of *Sarcocystis* spp. in Seabirds from Southern Brazil. *Parasitol. Int.* **2022**, *90*, 102595. [CrossRef] [PubMed]
62. Kutkienė, L.; Sruoga, A. *Sarcocystis* spp. in Birds of the Order Anseriformes. *Parasitol. Res.* **2004**, *92*, 171–172. [CrossRef]
63. Munday, B.L.; Hartley, W.J.; Harrigan, K.E.; Presidente, P.J.; Obendorf, D.L. *Sarcocystis* and related organisms in Australian wildlife: II. Survey findings in birds, reptiles, amphibians and fish. *J. Wildl. Dis.* **1979**, *15*, 57–73. [CrossRef]
64. Pak, S.M.; Eshtokina, V.N. Sarcosporidians of Birds. In *Sarcosporidians of Animals in Kazakhstan*; Panin, J.V., Ed.; Nauka: Alma-Ata, Kazakhstan, 1984.
65. Hirschfeld, A.; Heyd, A. Mortality of Migratory Birds caused by Hunting in Europe: Bag Statistics and Proposals for the Conservation of Birds and Animal Welfare. *Berichte Vogelschutz* **2005**, *42*, 47–74.
66. Wetlands International. Waterbird Population Estimates. 2020. Available online: www.wpe.wetlands.org (accessed on 7 April 2023).
67. Bright, P.W. Lessons from Lean Beasts: Conservation Biology of the Mustelids. *Mamm. Rev.* **2000**, *30*, 217–226. [CrossRef]
68. Zschille, J.; Stier, N.; Roth, M.; Mayer, R. Feeding Habits of Invasive American mink (*Neovison vison*) in Northern Germany—Potential Implications for Fishery and Waterfowl. *Acta Theriol.* **2013**, *59*, 25–34. [CrossRef]
69. Švažas, S.; Meissner, W.; Serebryakov, V.; Kozulin, A.; Grishanov, G. *Changes of Wintering Sites of Waterfowl in Central and Eastern Europe*; OMPO Vilnius & Institute of Ecology: Vilnius, Lithuania, 2001.

70. Švažas, S.; Kozulin, A. *Waterbirds of Large Fishponds of Belarus and Lithuania*; OMPO Vilnius & Institute of Ecology: Vilnius, Lithuania, 2002.
71. Viksne, J.; Švažas, S.; Czajkowski, A.; Janaus, M.; Mischenko, A.; Kozulin, A.; Kuresoo, A.; Serebryakov, V. *Atlas of Duck Populations in Europe*; Akstis Press: Vilnius, Lithuania, 2010.
72. Tuggle, B.N. Sarcocystis. In *Field Guide to Wildlife Diseases-General Field Procedures and Diseases of Migratory Birds*; Friend, M., Ed.; Resource Publication 167; United States Fish and Wildlife Service: Washington, DC, USA, 1987.
73. Andrew, R.; Newman, P.P. Sarcosporidiosis. *J. La. State Med. Soc.* **1961**, *113*, 224–226.
74. Keawcharoen, J.; Van Riel, D.; Van Amerongen, G.; Bestebroer, T.; Beyer, W.; Van Lavieren, R.; Osterhaus, A.; Fouchier, R.; Kuiken, T. Wild Ducks as Long-Distance Vectors of highly Pathogenic Avian Influenza Virus (H5N1). *Emerg. Infect. Dis.* **2008**, *14*, 600–607. [CrossRef] [PubMed]
75. Söderquist, P. Large-Scale Releases of Native Species: The Mallard as a Predictive Model System. Ph.D. Thesis, Swedish University of Agricultural Sciences, Uppsala, Sweden, 2015.

animals

MDPI

Article

Data on New Intermediate and Accidental Hosts Naturally Infected with *Angiostrongylus cantonensis* in La Gomera and Gran Canaria (Canary Islands, Spain)

Natalia Martin-Carrillo [1,2,3], Edgar Baz-González [1,2,3], Katherine García-Livia [1,2,4], Virginia Amaro-Ramos [1,2,3], Néstor Abreu-Acosta [1,5], Jordi Miquel [6,7], Estefanía Abreu-Yanes [1,5], Román Pino-Vera [1,2,3], Carlos Feliu [6,7] and Pilar Foronda [1,2,*]

1 Instituto Universitario de Enfermedades Tropicales y Salud Pública de Canarias (IUETSPC), Universidad de La Laguna (ULL), Avda. Astrofísico F. Sánchez, s/n, 38203 San Cristóbal de La Laguna, Canary Islands, Spain; nmartinc@ull.edu.es (N.M.-C.); ebazgonz@ull.edu.es (E.B.-G.); kgarcial@ull.edu.es (K.G.-L.); vamarora@ull.edu.es (V.A.-R.); gerencia@nertalab.es (N.A.-A.); esabya@gmail.com (E.A.-Y.); rpinover@ull.edu.es (R.P.-V.)
2 Department of Obstetricia y Ginecología, Pediatría, Medicina Preventiva y Salud Pública, Toxicología, Medicina Legal y Forense y Parasitología, Universidad de La Laguna (ULL), Avda. Astrofísico F. Sánchez, s/n, 38203 San Cristóbal de La Laguna, Canary Islands, Spain
3 Programa de Doctorado Ciencias Médicas y Farmacéuticas, Desarrollo y Calidad de Vida, Universidad de La Laguna (ULL), Avda. Astrofísico F. Sánchez, s/n, 38203 San Cristóbal de La Laguna, Canary Islands, Spain
4 Programa de Doctorado en Biodiversidad y Conservación, Universidad de La Laguna (ULL), Avda. Astrofísico F. Sánchez, s/n, 38203 San Cristóbal de La Laguna, Canary Islands, Spain
5 Nertalab S.L. José Rodríguez Mouré, 4, Bajo, 38008 Santa Cruz de Tenerife, Canary Islands, Spain
6 Departament de Biologia, Sanitat i Medi Ambient, Facultat de Farmàcia i Ciències de l'Alimentació, Universitat de Barcelona, Av. Joan XXIII, s/n, 08028 Barcelona, Ibérian Peninsula, Spain; jordimiquel@ub.edu (J.M.); cfeliu@ub.edu (C.F.)
7 Institut de Recerca de la Biodiversitat (IRBio), University of Barcelona, Av Diagonal, 645, 08028 Barcelona, Ibérian Peninsula, Spain
* Correspondence: pforonda@ull.edu.es

Simple Summary: The rat lungworm, *Angiostrongylus cantonensis*, is the world's leading cause of eosinophilic meningitis in humans. It is an emerging zoonotic parasite endemic to Asia and the Pacific Islands that has spread to all continents except Antarctica. In the Canary Islands, its presence was first detected in 2010 on the island of Tenerife. Numerous studies demonstrated the high capacity of *A. cantonensis* to colonize new areas, especially its ability to parasitize a wide range of animals. Due to the similarity of the ecosystems that we can find between the islands that make up the Canary Archipelago and the great diversity of species of both vertebrates and terrestrial gastropods, the objective of this study was to analyze several species as possible hosts of *A. cantonensis* on other islands in the Canary Islands, with the use of molecular tools. The present study confirmed the presence of *A. cantonensis* in two mammalian species, *Mus musculus* and *Felis catus*, and in four terrestrial gastropod species, *Limacus flavus, Milax gagates, Insulivitrina emmersoni*, and *Insulivitrina oromii*. The presence of *A. cantonensis* implies a possible risk to humans and other animals, which justifies the need for control measures to control the possible risk of infection and, thus, prevent public health and veterinary problems.

Abstract: *Angiostrongylus cantonensis* is a metastrongyloid nematode and the etiologic agent of angiostrongyliasis, a disease characterized by eosinophilic meningitis. This emerging zoonotic parasite has undergone great expansion, including in some regions of Europe and America. In the Canary Islands, the parasite was first discovered parasitizing *Rattus rattus* on the island of Tenerife in 2010. To date, the distribution of this parasite in the Canary Islands has been restricted to the northern zone and the main cities of Tenerife. Using molecular tools for the sentinel species present in the Canary Islands, this study confirmed the presence of the nematode on two other islands in the Canary Archipelago: La Gomera and Gran Canaria. Furthermore, this emerging parasite was detected, besides in the common definitive host *R. rattus*, in wild *Mus musculus* and *Felis catus* and in four terrestrial gastropod species, *Limacus flavus, Milax gagates, Insulivitrina emmersoni,* and *Insulivitrina oromii*, two of them endemic to La Gomera, for the first time, increasing the number of non-definitive host species. This study reinforces the expansion character of *A. cantonensis* and highlights the importance of knowledge about sentinel species for identifying new transmission locations that help prevent and control the transmission of the parasite and, thus, prevent public health problems.

Keywords: *Angiostrongylus cantonensis*; eosinophilic meningitis; *Rattus rattus*; *Mus musculus*; *Felis catus*; *Limacus flavus*; *Milax gagates*; *Insulivitrina oromii*; *Insulivitrina emmersoni*

1. Introduction

The rat lungworm, *Angiostrongylus cantonensis*, is a parasitic nematode that causes a reemerging zoonotic disease, human eosinophilic meningitis, as well as neurological abnormalities in wild and domestic animals [1,2]. It has been reported worldwide since its first description in the Guangzhou region of China in the pulmonary arteries of brown rats, *Rattus norvegicus* [3]. This parasite has a complex life cycle involving various species of rats as definitive hosts [4], numerous gastropod species from widely different taxonomic groups as intermediate hosts, and various paratenic hosts. It was identified as a human pathogen in 1945 [5] and is now recognized as the leading cause of eosinophilic meningitis worldwide [6]. The most recent globally reported cases of human neuroangiostrongyliasis were approximately 7000 cases [7]. These data may be higher since many cases are not published. As non-permissive hosts, humans mainly become infected by eating raw or undercooked intermediate or paratenic hosts containing the infective larvae and by eating vegetables and salads contaminated with this parasite [7,8].

Upon ingestion, the infective larvae invade the intestinal wall, causing enteritis, and enter the bloodstream [9]. The central nervous system (CNS) is the most common site of migration and is the main clinical manifestation of eosinophilic meningitis. Human angiostrongyliasis presents with a broad clinical spectrum, ranging from mild disease to eosinophilic meningitis or encephalitis [10]. As a result, neurological damage and even death may occur, especially if prompt and proper treatment is not administered [11–13].

Traditionally, *A. cantonensis* was an endemic parasite of the tropical and subtropical areas of the Far East, including Southeast Asia, the Pacific Islands, areas of South and Central America, and the Caribbean [9,14]. It was widely spread, mainly across the tropics and subtropics [15]. More recently, it was reported in rodents (*Rattus rattus*) on the Atlantic archipelago of the Canary Islands (Spain) [16]; on the Mediterranean island of Mallorca (Spain) parasitizing hedgehog, *Atelerix algirus* [17]; and later in the rodent populations (*R. rattus* and *R. norvegicus*) of continental Europe (Valencia, Spain) [18]. These data suggest that despite the lack of tolerance of *A. cantonensis*, which has traditionally been largely limited to tropical and subtropical regions, to cold temperatures [19], with the fulcrum of its presence in Europe, the parasite could spread across the continent to more temperate regions [15], as already mentioned in Australia [20] and the United States [21]. These data suggest the geographic expansion of the parasite, resulting in a possible rapid increase in the incidence of infection in humans [22].

The influence of increasing factors, such as globalization and climate change, may influence the invasion of non-endemic regions by lungworms. Furthermore, many authors attributed the spread of *A. cantonensis* to the great diversity of its intermediate hosts and its high adaptability to new intermediate host species [21]. In addition, the invasion of *A. cantonensis* was associated with cargo shipments by aircraft or ships, which sometimes unintentionally import rats and gastropods [9,23].

The Canary Islands constitute an archipelago consisting of eight volcanic islands located in the Atlantic Ocean that settled on the African plate. This archipelago is in the

Northwest of the African Coast near Southern Morocco and northern Western Sahara (13°23–18°8 W and 27°37–29°24 N). It is one of the outermost regions of the European Union and a part of the natural region of Macaronesia, of which it is the largest and most populated archipelago. The population of the Canary Archipelago is mainly concentrated on its two capital islands, Tenerife and Gran Canaria, at approximately 43% and 40%, respectively [24]. In the Canary Islands, *A. cantonensis* was detected in *R. rattus* on the island of Tenerife in 2010 [16] and posteriorly in three mollusk species [25] and in the endemic lizards *Gallotia galloti* [26]. Although previous studies were carried out with rodents from all the Canary Islands, *A. cantonensis* was only detected in Tenerife [23].

Taking into account the rapid dispersal of *A. cantonensis* worldwide and the health implications for humans and wildlife, there is a need to determine the distribution of *A. cantonensis* in the Canary Archipelago. Given the similarity of the climatic and orographic conditions between Tenerife and the other islands of the Canary Archipelago and the fact that *A. cantonensis* was reported in non-definitive hosts, such as *A. algirus*, in other regions [17], the aim of this study was to detect the presence of *A. cantonensis* using molecular tools in different possible host species on two other Canary Islands, La Gomera and Gran Canaria.

2. Materials and Methods

2.1. Samples Collection

From February 2022 to January 2023, sampling campaigns were conducted in rural and urban areas on La Gomera and Gran Canaria islands (Figure 1). On the island of La Gomera, rural areas are mainly composed of Laurisilva and Fayal-Brezal zones, biotopes with the presence of horizontal rain biotopes. On Gran Canaria, rural areas are composed of cultivated fields, ravines, and pine forests.

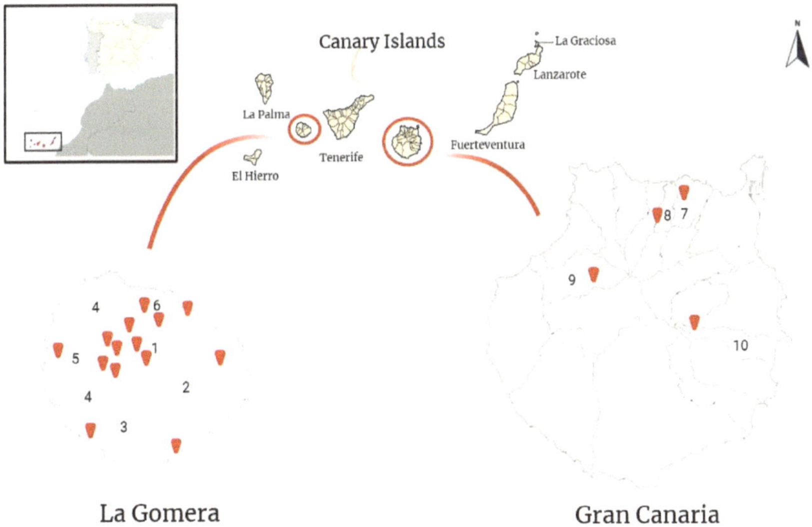

Figure 1. Map of the sampled areas on the islands of La Gomera and Gran Canaria (Canary Islands, Spain). In the upper left corner, the geographical situation of the Canary Islands is shown in relation to the African continent and the Iberian Peninsula. The red marks show the sampled locations. The numbers indicate the sampled municipalities. 1: Hermigua; 2: San Sebastian de La Gomera; 3: Alajeró; 4: Vallehermoso; 5: Valle Gran Rey; 6: Agulo; 7: Arucas; 8: Firgas; 9: Artenara; 10: Ingenio. (Image created with BioRender.com; accessed on 30 March 2023).

Samples of rodents of *Mus musculus* (60), *R. norvegicus* (4), and *R. rattus* (40) species were obtained from six and four municipalities of La Gomera and Gran Canaria islands, respectively (Figure 1 and Tables 1 and 2), using live traps. Traps were set in the afternoon and collected the following day in the morning. Once captured, euthanasia was performed with CO_2 or by cervical dislocation for rats and mice, respectively. Animal work was approved and authorized by "Consejería de Transición Ecológica, Lucha contra el Cambio Climático y Planificación Territorial" (Gobierno de Canarias) (Expte. 2021/51732).

Table 1. Origin of analyzed host species in this study.

Host Species	Island	Sample Size	Total
Felis catus	La Gomera	40	40
Mus musculus	La Gomera Gran Canaria	34 26	60
Rattus norvegicus	La Gomera	4	4
Rattus rattus	La Gomera Gran Canaria	33 7	40
Insulivitrina oromii	La Gomera	8	8
Insulivitrina emmersoni	La Gomera	28	28
Limacus flavus	La Gomera	10	10
Milax gagates	Gran Canaria	24	24
		Total	214

Table 2. Prevalence of *Angiostrongylus cantonensis* obtained on the La Gomera and Gran Canaria islands (Canary Islands, Spain) in this study, divided by host species as well as location.

Island	Location	Host Species	Prevalence of *Angiostrongylus cantonensis* P (%) * (+/n)	Total P (%) * (+/n)
La Gomera	Agulo	*Felis catus* *Mus musculus* *Insulivitrina oromii*	0 (0/2) 0 (0/5) 50 (1/2)	11.11 (1/9)
	Alajeró	*Felis catus* *Mus musculus* *Insulivitrina oromii*	0 (0/1) 0 (0/3) 0 (0/2)	0 (0/6)
	Hermigua	*Felis catus* *Mus muculus* *Insulivitrina oromii* *Limacus flavus*	12.5 (1/8) 0 (0/15) 66.06 (2/3) 60 (6/10)	25% (9/36)
	San Sebastián de la Gomera	*Mus musculus*	0 (0/6)	0 (0/6)
	Valle Gran Rey	*Felis catus* *Insulivitrina oromii*	0 (0/6) 0 (0/1)	0 (0/7)
	Vallehermoso	*Felis catus* *Mus musculus* *Insulivitrina emmersoni*	17.39 (4/23) 0 (0/5) 32.14 (9/28)	23.21 (13/56)
Total of La Gomera				19.16 (23/120)
Gran Canaria	Artenara	*Mus musculus*	0 (0/2)	0 (0/2)
	Arucas	*Mus musculus*	16.66 (1/6)	16.66 (1/6)
	Firgas	*Mus musculus* *Milax gagates*	100 (1/1) 4.16 (1/24)	8 (2/25)

Table 2. *Cont.*

Island	Location	Host Species	Prevalence of *Angiostrongylus cantonensis* P (%) * (+/n)	Total P (%) * (+/n)
	Ingenio	*Mus musculus*	41.17 (7/17)	41.17 (7/17)
Total of Gran Canaria				20 (10/50)
Total				13.52% (33/170)

* Prevalence of *Angiostrongylus cantonensis* %; (+/n): positive samples for nested PCR/analyzed samples.

Further, terrestrial gastropods were collected from five municipalities in La Gomera and one in Gran Canaria (Figure 2 and Tables 1 and 2). The collected terrestrial gastropods included the endemic species *Insulivitrina emmersoni* (28) and *Insulivitrina oromii* (8), present only on the island of La Gomera (Figure 2 and Tables 1 and 2). On the other hand, *Limacus flavus* (10) and *Milax gagates* (24), possible native species to the Canary Islands, were collected on the islands of La Gomera and Gran Canaria, respectively (Figure 2 and Tables 1 and 2). This field study was authorized by the "Consejería de Transición Ecológica, Lucha contra el Cambio Climático y Planificación Territorial" (Gobierno de Canarias) (Expte. 19-2022/1110092705). The capture of animals in protected areas (Garajonay National Park) was approved by "Cabildo Insular de La Gomera" (Expte. 4121/2022) and "Parque Nacional de Garajonay" (749.153, TELP/122.918).

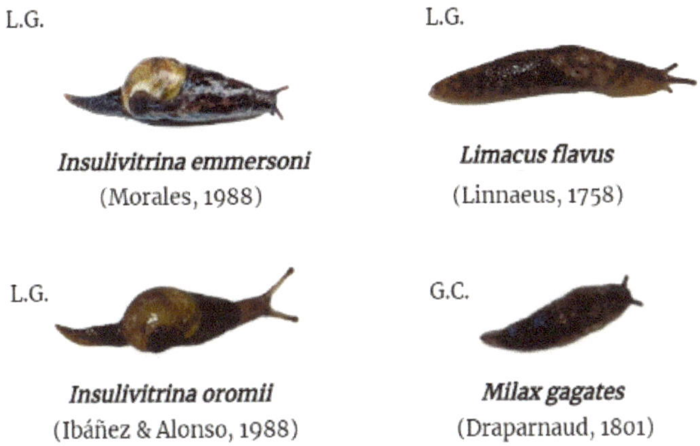

Figure 2. Terrestrial gastropod species included in the study. L.G. collected on the island of La Gomera; G.C. collected on the island of Gran Canaria.

On the other hand, *Felis catus* (40) samples from La Gomera were also analyzed (Tables 1 and 2). These animals were donated for this study by Gobierno de Canarias staff (approved by the Gobierno de Canarias, expedient number 2022/25073, and Excmo. Cabildo Insular de La Gomera, expedient numbers 1872/2021 and 1821/2022).

2.2. *Search for Angiostrongylus cantonensis and Sample Isolation*

After dissection of rodents, the lungs and hearts were isolated in physiological saline and examined with a Leica M80 stereo microscope (Leica Mikrosysteme Vertrieb GmbH, Wetzlar, Germany) to detect the presence of *A. cantonensis*. For terrestrial gastropods, tissue samples were placed in 1 mL of 0.01% pepsin0.7% HCl for digestion and observed under the Leica DM750 Microscopium model ICC50 HD (Leica Microsystems, Heerbrugg, Switzerland) for the search for larval forms of *A. cantonensis*.

On the other hand, for rodents and *F. catus*, approximately 1 g of the liver was aseptically removed using a sterile scalpel. For terrestrial gastropods, samples were cut from the posterior end of the mollusk foot (~0.5 g) under aseptic conditions. The samples were fixed in absolute ethanol and stored at 4 °C until DNA extraction.

2.3. DNA Extraction

Genomic DNA was extracted using approximately 25 mg of liver tissue, and the posterior end of terrestrial gastropod foot tissue was cut into small pieces and used for DNA extraction. Other tissues were not included in the analyses, following the success of Anettová et al. [26] looking for DNA of *A. cantonensis* in the livers of paratenic hosts. The tissues were deposited in tubes containing a mixture of 250 μL of a lysis solution composed of 30 mM Tris-HCL pH 8.0 mM EDTA and 0.4% SDS, and three microliters of proteinase K (20 mL^{-1}) (PanReac AppliChem ITW Reagents, Germany) was added and incubated at 56 °C overnight. Then, 250 μL of NH_4Ac 4M was added, thoroughly mixed, and subsequently incubated for 30 min at room temperature (15–25 °C). The mix was spun for 10 min at 13,000 rpm, and the pellet was discarded. DNA was then precipitated from the supernatant with ethanol, and the pellet was suspended in 200 μL of 1X TE (10 mM Tris-HCL pH 8.1 mM EDTA) [27]. The quantity and quality of genomic DNA were checked using DeNovix DS-11 + Spectrophotometer (DeNovix Inc., Wilmington, DE, USA) (Table S1).

2.4. PCR Amplification

For the detection of *A. cantonensis* in the extracted genomic DNA, nested PCR was performed using the primers (10 μM) described by Qvarnstrom et al. [28] to amplify the entire region of internal spacer 1 (ITS1) and the second PCR with specific primers for *A. cantonensis*, as described by Izquierdo-Rodríguez et al. [29]. Both rounds of PCR contained 1X buffer (VWR International, Haasrode, Belgium), 0.2 mM of each dNTP (VWR International, Haasrode, Belgium), 1.5 mM $MgCl_2$ (VWR International, Haasrode, Belgium), and 20–40 ng of total genomic DNA for each sample in a total volume of 25 μL. Positive controls (gDNA extracted from *A. cantonensis* adult worm with the procedure described by López et al. [27]) and negative controls (H_2O) were included in all assays to verify nested PCR performance. All DNA samples were run in duplicate. Amplification was performed with the XP Cycler (Bioer Technology, Hangzhou, China) using the following parameters: 94 °C for 5 min; 35 cycles at 94 °C for 30 s, 57 °C for 90 s, and 72 °C for 90 s; and a final extension at 72 °C for 10 min. The resulting amplifications were checked on 1.5% agarose gel.

2.5. Sequencing and Sequencing Data Analysis

PCR products with the expected size (642-bp) were sequenced at Macrogen Spain (Madrid, Spain) with primers ITS1 Canto F3 and ITS1 Canto R1 [29]. The sequences obtained using the Sanger method were interpreted with MEGA X software [30], using the multiple alignment program ClustalW included in MEGA X, and minor corrections were made by hand. Subsequently, the sequences were analyzed using the Basic Local Alignment Search Tool (BLAST), and their identities were confirmed by homology comparison.

Phylogenetic relationships based on the neighbor joining [31] method were carried out with the Kimura 2-parameter model [32], and 1000 bootstrap replications explored the relationships among them. The sequence of *Protostrongylus oryctolagi* (Acc. Number: OM307447) was used as the outgroup.

3. Results

After observation of the rodent lungs and hearts, the nematode *A. cantonensis* was not observed in any of the specimens. In the case of terrestrial gastropods, microscopic observation was negative for the presence of *A. cantonensis* larvae. However, it should be noted that some of the individuals had observed structures similar to the larvae of this parasite, but due to the poor conditions in which they were found, it was impossible to perform a morphometric study of them. Regarding the molecular study, *A. cantonensis* was detected on the two analyzed islands, La Gomera and Gran Canaria, and in all the analyzed species, both vertebrates and invertebrates.

The overall prevalence of *A. cantonesis* was 15% in *M. musculus* (9/60); 12.5% (5/40) in feral cats, and 27.14% in terrestrial gastropods (*I. oromii*: 37.5% (3/8); *I. emmersoni*: 32.14% (9/28); *L. flavus*: 60% (6/10); and *M. gagates*: 4.16% (1/24)). The prevalence data for all the analyzed hosts and areas are shown in Table 2.

The molecular results for the definitive hosts revealed the presence of *A. cantonensis* in the 2.5% (1/40) of *R. rattus* that were analyzed, concretely in the municipality of Hermigua (7.14%; 1/14) on La Gomera island.

The BLAST analysis confirmed that the sequences obtained from all the hosts analyzed in this study showed 100% homology with *A. cantonensis* (Query Cover (Q.C.): 100%). With respect to the other Angiostrongylidae species, the BLAST analyses showed a lower identity: 84.8% with *Angiostrongylus vasorum* (Q.C.: 39%) and *Angiostrongylus chabaudi* (Q.C.: 40%) and 81% with *Angiostrongylus abstrusus* (Q.C.: 19%).

Of the 34 amplified samples obtained by nested PCR (Figure 3), 16 sequences were obtained for the 18 S ribosomal RNA gene partial sequence, the internal transcribed spacer 1 complete sequence, and the 5.8S ribosomal RNA gene partial sequence. Phylogenetic analyses confirmed the identity of *A. cantonensis* observed by BLAST with an average Query Cover and Identity of 99.8% and 99%, respectively.

Figure 3. Results of nested PCR for *Angiostrongylus cantonensis* (642 bp) in samples included in this study. Line 1: *Felis catus*; Lane 2: *Mus musculus*; Line 4: *Rattus rattus*; Line 7: *Insulivitrina emmersoni*; Line 8: *Insulivitrina oromii*; Line 10: *Limacus flavus*; Line 11: Milex gagates; Line 19: negative control; Line 20: positive control; Lines 3, 5, 6, and 12–18: negative samples for *A. cantonensis* DNA; Line 9: molecular ladder ranging from 100 to 4000 bp (VWR International, Haasrode, Belgium). (1*, 2*, and 3* each refer to the above lines.)

The nucleotide sequences obtained in this study were submitted to the GenBank database under accession numbers OQ702304–OQ702319. An alignment of 482 bp was used for phylogenetic analysis. The results of the neighbor joining analysis based on the partial sequence of internal transcribed spacer 1 are shown in Figure 4.

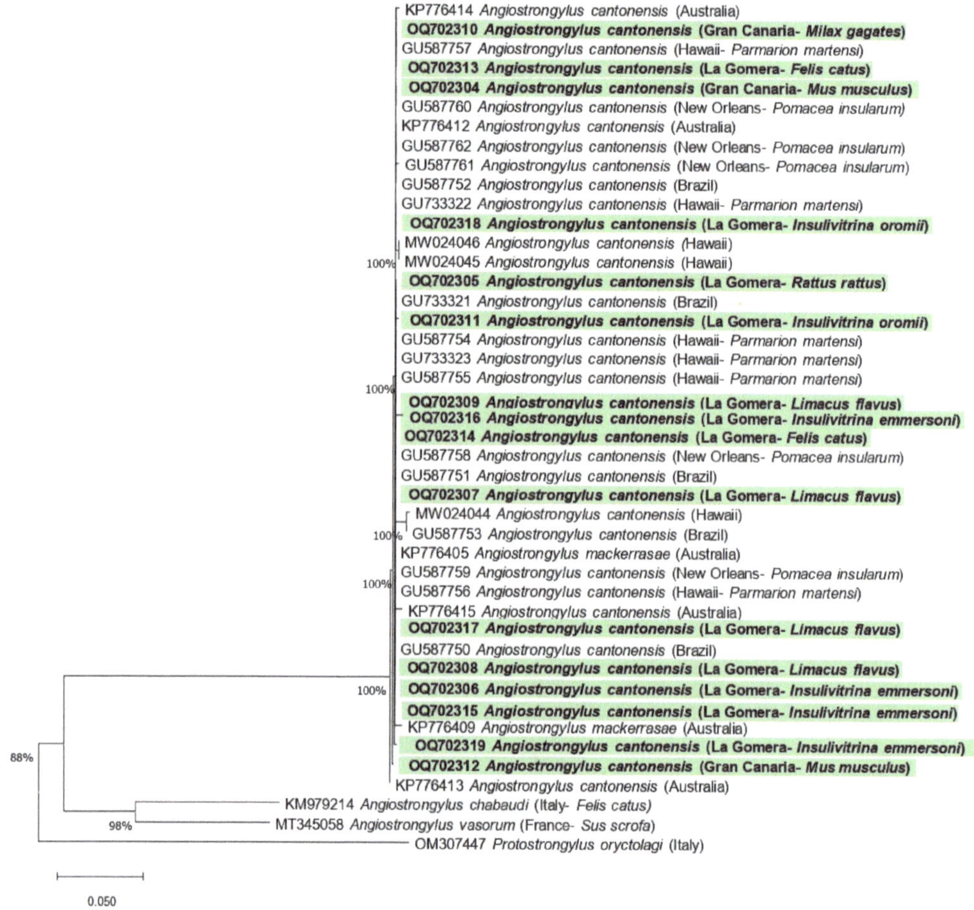

Figure 4. Phylogenetic relationships between sequences of the partial sequence of the internal transcribed spacer 1 of *Angiostrongylus* spp., including the nucleotide sequences obtained in this study (shown in bold and highlighted in green). The tree was built using neighbor joining method based on genetic distance calculated by the Kimura 2-parameter model. There were a total of 482 positions in the final data set.

4. Discussion

The present study constitutes the first detection of *A. cantonensis* on the islands of La Gomera and Gran Canaria (Canary Islands, Spain). Previous studies reported this nematode on the neighboring island of Tenerife, but it was not detected on any other Canary Islands [16,23]. Therefore, the present results demonstrate the expansion of *A. cantonensis* to new locations in the Canary Archipelago, confirming the colonizer character of this nematode, recently widely reported by other authors, such as Barrat et al. [22], and Cowie et al. [15], among others. In the present work, on La Gomera and Gran Canaria, *A. cantonensis* was detected by molecular methods both in mammals and invertebrates; specifically, in two of the analyzed rodent species, *M. musculus* and *R. rattus*; in *F. catus*; and in the four analyzed terrestrial gastropod species, *I. emmersoni*, *I. oromii*, *L. flavus*, and *M. gagates*; all of these are the first reports, except for *R. rattus*, of natural infection in these host species.

The role of many rodent species as definitive hosts of *A. cantonensis* is widely known [22]. These mammals are present in different ecosystems that are widely distributed throughout the world. In this study, we included the species *M. musculus* and *F. catus* on the list of accidental hosts, which are also present worldwide. Both species were introduced to the Canary Islands and are present on the eight islands that make up the Canary Archipelago.

Despite the numerous studies carried out on *A. cantonensis* in *M. musculus*, it was only reported in experimentally infected mice [33,34]. With the detection of genetic material of *A. cantonensis* in wild *M. musculus*, we can confirm the ability of this vertebrate to be naturally infected by *A. cantonensis*, increasing the list of the accidental hosts of this parasite. This host may also play a fundamental role in the dispersion to non-endemic regions.

Regarding *F. catus*, there are no previous data about its parasitization by *A. cantonensis*. In the Canary Islands, the feral cat is widely distributed on all the islands, being a top predator in the vertebrate food chain [35]. Its role as an accidental host of *A. cantonensis* is now confirmed with the detection of its DNA. This fact makes sense considering that its diet includes mammals, birds, reptiles, gastropods, and arthropods [36,37], many of which were confirmed as hosts of *A. cantonensis* in the Canary Islands, as is the case of the endemic lizard *G. galloti*, recently cited as a paratenic host [26]. It is important to note that the presence of *A. cantonensis* in feral cats does not involve health risks for humans.

Regarding the matter of the studied gastropod species, the four analyzed species presented DNA of *A. cantonensis*, also increasing the number of known intermediate hosts. *Limacus flavus* and *M. gagates* are common species with a wide distribution in the Canary Islands and are also widely distributed in Europe, Asia, and South America [38]. However, *I. oromii* and *I. emmersoni* are endemic to the Canary Islands and are exclusively found on the island of La Gomera in the biotopes of Laurisilva and Monteverde [38]. This finding is consistent with the previously demonstrated ability of *A. cantonensis* to naturally infect a wide variety of freshwater and terrestrial gastropods species [25,39].

The biotopes where *A. cantonensis* was detected coincides with the biotopes where this parasite had previously been detected on the island of Tenerife [23]. Specifically, on the island of La Gomera, the areas where it is located coincide with the areas of incidence of trade winds and horizontal rain, and on the island of Gran Canaria the areas where it is located coincide with the urban and peri-urban areas.

Some studies relate the rate of development of *A. cantonensis* in intermediate hosts or sentinel species to climate [40]. The survival of many of the hosts of this parasite depends on optimal humidity and temperature conditions for their development and life cycle, as can be the case for many terrestrial gastropods. Overall, in most natural systems, climate change affects hosts and parasites by altering their survival, reproduction, and transmission, among other factors [21,41,42]. Additionally, increased international trade, travel to disease-endemic regions, and changes in dietary habits are collectively expected to facilitate the further range expansion of *A. cantonensis* [43,44]. This fact was verified by numerous studies citing this emerging zoonotic parasite in many regions of the world under diverse environmental conditions [15,21].

The current rapid expansion of a multitude of emerging pathogens indicates that the control levels of many of them must be increased to avoid public health problems, both locally and globally. Many studies advocate the use of sentinel animals because of their potential [45], and, by studying a wide range of potential hosts, we can increase vigilance regarding the spread of emerging pathogens such as *A. cantonensis*.

Autochthonous or introduced species that act as hosts in regions recently invaded by *A. cantonensis* were previously identified [41,45,46], as in this study. Hence, further research is needed to confirm the presence of the rat lungworm in a wide range of animals that can act as sentinel species and even as intermediate or paratenic hosts. One factor that should be considered for any future epidemiological studies of this emerging pathogen is the technique that we used to detect its presence, since, as we were able to verify in this study, it is not always possible to visualize the adult in definitive hosts. Therefore, the use of more sensitive techniques is vital to rule out false negatives in species with the potential to harbor the parasite.

In many regions, awareness of the rat lungworm is low, not only among the general community but also among physicians and health authorities. Therefore, the education of medical professionals and local and tourist populations must be prioritized in endemic areas. This makes even more sense if we observe that the spread of this emerging zoonotic pathogen has exponentially increased over the years.

5. Conclusions

Overall, the results of this study indicate, for the first time, the presence of *A. cantonensis* on La Gomera and Gran Canaria, in the Canary Islands, using molecular tools. In addition, the parasite was first reported in naturally infected *M. musculus* and *F. catus* and in four species of terrestrial gastropods: *I. emmersoni, I. oromomii, L. flavus*, and *M. gagates*. All of these have a wide distribution, except for the gastropod species *I. emmersoni* and *I. oromii*, which are endemic to La Gomera. Based on these data, this study reinforces the high potential for the expansion of *A. cantonensis* and stresses the importance of the knowledge that sentinel species have when controlling infectious or parasitic diseases to indicate possible sources of risk of infection.

Supplementary Materials: The following supporting information can be downloaded at: https://www.mdpi.com/article/10.3390/ani13121969/s1, Table S1: Quantification of the DNA obtained in the study samples.

Author Contributions: Conceptualization, P.F.; methodology, N.M.-C., E.B.-G., K.G.-L., V.A.-R., N.A.-A., J.M., E.A.-Y., R.P.-V., C.F. and P.F.; formal analysis, N.M.-C., E.B.-G., K.G.-L. and P.F.; investigation, N.M.-C.; resources, P.F.; data curation, N.M.-C., E.B.-G., K.G.-L., N.A.-A. and P.F.; writing—original draft preparation, N.M.-C. and P.F.; writing—review and editing, N.M.-C., E.B.-G., K.G.-L. and P.F.; visualization, P.F.; supervision, P.F.; project administration, P.F.; funding acquisition, P.F. All authors have read and agreed to the published version of the manuscript.

Funding: This research was funded by ProID2021010013 (Consejería de Economía, Industria, Comercio y Conocimiento, Gobierno de Canarias), Fondos FEDER-FSE Canarias 2014–2020, "Programa de Apoyo a la Investigación María del Carmen Betencourt y Molina". This study was also supported by the "Consejería de Transición Ecológica, Lucha contra el Cambio Climático y Planificación Territorial" (Gobierno de Canarias)–Universidad de La Laguna agreement "Estudio de patógenos en aves migratorias y en especies exóticas en un escenario de cambio climático". E.B.-G. and R.P.-V. were granted FPI predoctoral scholarships by the "Agencia Canaria de Investigación, Innovación y Sociedad de la Información de la Consejería de Economía, Conocimiento y Empleo" and the "Fondo Social Europeo (FSE) Programa Operativo Integrado de Canarias 2014–2020, Eje 3 Tema Prioritario 74 (85%)" (TESIS2021010056 and TESIS2022010038). K.G.L. was granted a scholarship by the Ministerio de Ciencia, Innovación y Universidades de España, and Universidad de La Laguna (Becas M-ULL, convocatoria 2019).

Institutional Review Board Statement: The capture of rodents was approved and authorized by "Ministerio de Transición Ecológica, Lucha contra el Cambio Climático y Planificación Territorial" (Gobierno de Canarias) (Expte. 2021/51732). This study did not involve protected species or species in danger; the capture was authorized by the "Ministerio de Transición Ecológica, Lucha contra el Cambio Climático y Planificación Territorial" (Gobierno de Canarias) (Expte. 19-2022/1110092705). The capture of animals in protected areas (Parque Nacional de Garajonay) was approved by the "Cabildo Insular de La Gomera" (Expte. 4121/2022) and "Parque Nacional de Garajonay" (749.153, TELP/122.918). Finally, the specimens of *Felis catus* were donated for this study, already sacrificed, with the authorization of the "Dirección General de Lucha contra el Cambio Climático y el Medio Ambiente" (Gobierno de Canarias) (Expte. 2021/51732).

Informed Consent Statement: Not applicable.

Data Availability Statement: Not applicable.

Acknowledgments: We would like to thank the staff of RedEXOS and Gestión y Planeamiento Territorial y Medioambiental, S.A. (GESPLAN) for donating animals; Leopoldo Moro Abad for the identification of the terrestrial gastropods. J.M. is a member of the 2021-SGR-00359 research group.

Conflicts of Interest: The authors declare no conflict of interest.

References

1. Kim, D.Y.; Stewart, T.B.; Bauer, R.W.; Mitchell, M. *Parastrongylus* (=*Angiostrongylus*) *cantonensis* now endemic in Louisiana wildlife. *J. Parasitol.* **2002**, *88*, 1024–1026. [CrossRef] [PubMed]
2. Qvarnstrom, Y.; Sullivan, J.J.; Bishop, H.S.; Hollingsworth, R.; da Silva, A.J. PCR-based detection of *Angiostrongylus cantonensis* in tissue and mucus secretions from molluscan hosts. *Appl. Environ. Microbiol.* **2007**, *73*, 1415–1419. [CrossRef]
3. Chen, H.T. Un nouveau nématode pulmonaire, *Pulmonema cantonensis*, ng, n. sp. *Ann. Parasitol. Hum. Comp.* **1935**, *13*, 312–317. [CrossRef]
4. Yong, H.S.; Eamsobhana, P. Definitive rodent hosts of the rat lungworm *Angiostrongylus cantonensis. Raffles Bull. Zool.* **2013**, *29*, 111–115.
5. Beaver, P.C.; Rosn, L. Memorandum on the First Report of *Angiostrongylus* in Man, by Nomura and Lin, 1945. *Am. J. Trop. Med. Hyg.* **1964**, *13*, 589–590. [CrossRef] [PubMed]
6. Murphy, G.S.; Johnson, S. Clinical Aspects of Eosinophilic Meningitis and Meningoencephalitis caused by Angiostrongylus cantonensis, the Rat Lungworm. *Hawaii J. Med. Public Health* **2013**, *72*, 35.
7. Turck, H.C.; Fox, M.T.; Cowie, R.H. Paratenic hosts of *Angiostrongylus cantonensis* and their relation to human neuroangiostrongyliasis globally. *One Health* **2022**, *15*, 100426. [CrossRef] [PubMed]
8. Wang, H.; Lu, L.; She, D.; Wen, Z.; Mo, Z.; Li, J.; Li, H. Eating Centipedes Can Result in Angiostrongylus cantonensis Infection: Two Case Reports and Pathogen Investigation. *Am. J. Trop. Med. Hyg.* **2018**, *99*, 743. [CrossRef] [PubMed]
9. Wang, Q.-P.; Lai, D.-H.; Zhu, X.-Q.; Chen, X.-G.; Lun, Z.-R. Human angiostrongyliasis. *Lancet Infect. Dis.* **2008**, *8*, 621–630. [CrossRef]
10. Slom, T.J.; Cortese, M.M.; Gerber, S.I.; Jones, R.C.; Holtz, T.H.; Lopez, A.S.; Zambrano, C.H.; Sufit, R.L.; Sakilvaree, Y.; Chaicumpa, W.; et al. An outbreak of eosinophilic meningitis caused by *Angiostrongylus cantonensis* in travelers returning from the Caribbean. *N. Engl. J. Med.* **2002**, *346*, 668–675. [CrossRef] [PubMed]
11. Yii, C.-Y. Clinical Observations on Eosinophilic Meningitis and Meningoencephalitis caused by *Angiostrongylus cantonensis* on Taiwan. *Am. J. Trop. Med. Hyg.* **1976**, *25*, 233–249. [CrossRef]
12. Bowden, D.K. Eosinophilic Meningitis in the New Hebrides: Two Outbreaks and two Deaths. *Am. J. Trop. Med. Hyg.* **1981**, *30*, 1141–1143. [CrossRef] [PubMed]
13. Chotmongkol, V.; Sawanyawisuth, K. Clinical manifestations and outcome of patients with severe eosinophilic meningoencephalitis presumably caused by *Angiostrongylus cantonensis*. *Age* **2002**, *45*, 30–61.
14. Cowie, R.H. Biology, Systematics, Life Cycle, and Distribution of *Angiostrongylus cantonensis*, the Cause of Rat Lungworm Disease. *Hawaii J. Med. Public Health* **2013**, *72*, 6. [PubMed]
15. Cowie, R.H.; Ansdell, V.; Panosian Dunavan, C.; Rollins, R.L. Neuroangiostrongyliasis: Global Spread of an Emerging Tropical Disease. *Am. J. Trop. Med. Hyg.* **2022**, *107*, 1166–1172. [CrossRef] [PubMed]
16. Foronda, P.; López-González, M.; Miquel, J.; Torres, J.; Segovia, M.; Abreu-Acosta, N.; Casanova, J.C.; Valladares, B.; Mas-Coma, S.; Bargues, M.D.; et al. Finding of *Parastrongylus cantonensis* (Chen, 1935) in Rattus rattus in Tenerife, Canary Islands (Spain). *Acta Trop.* **2010**, *114*, 123–127. [CrossRef]
17. Paredes-Esquivel, C.; Sola, J.; Delgado-Serra, S.; Puig Riera, M.; Negre, N.; Miranda, M.Á.; Jurado-Rivera, J.A. Angiostrongylus cantonensis in North African hedgehogs as vertebrate hosts, Mallorca, Spain, October 2018. *Eurosurveillance* **2019**, *24*, 1900489. [CrossRef]
18. Galán-Puchades, M.T.; Gómez-Samblás, M.; Osuna, A.; Sáez-Durán, S.; Bueno-Marí, R.; Fuentes, M.V. Autochthonous *Angiostrongylus cantonensis* Lungworms in Urban Rats, Valencia, Spain, 2021. *Emerg. Infect. Dis.* **2022**, *28*, 2564–2567. [CrossRef]
19. Lv, S.; Zhang, Y.; Steinmann, P.; Yang, G.J.; Yang, K.; Zhou, X.N.; Utzinger, J. The emergence of angiostrongyliasis in the People's Republic of China: The interplay between invasive snails, climate change and transmission dynamics. *Fresh Biol.* **2011**, *56*, 717–734. [CrossRef]
20. Stokes, V.L.; Spratt, D.M.; Banks, P.B.; Pech, R.P.; Williams, R.L. Occurrence of Angiostrongylus species (Nematoda) in populations of Rattus rattus and Rattus fuscipes in coastal forests of south-eastern Australia. *Aust. J. Zool.* **2007**, *55*, 177–184. [CrossRef]
21. York, E.M.; Butler, C.J.; Lord, W.D. Global Decline in Suitable Habitat for *Angiostrongylus* (=*Parastrongylus*) *cantonensis*: The Role of Climate Change. *PLoS ONE* **2014**, *9*, e103831. [CrossRef] [PubMed]
22. Barratt, J.; Chan, D.; Sandaradura, I.; Malik, R.; Spielman, D.; Lee, R.; Lee, R.; Marriott, D.; Harkness, J.; Ellis, J.; et al. *Angiostrongylus cantonensis*: A review of its distribution, molecular biology and clinical significance as a human pathogen. *Parasitology* **2016**, *143*, 1087–1118. [CrossRef]
23. Martín-Carrillo, N.; Feliu, C.; Abreu-Acosta, N.; Izquierdo-Rodriguez, E.; Dorta-Guerra, R.; Miquel, J.; Abreu-Yanes, E.; Martín-Alonso, A.; García-Livia, K.A.; Quispe-Ricalde, M.A.; et al. A peculiar distribution of the emerging nematode *Angiostrongylus cantonensis* in the Canary Islands (Spain): Recent introduction or isolation effect? *Animals* **2021**, *11*, 1267. [CrossRef] [PubMed]

24. Instituto Canario de Estadística (ISTAC). Available online: http://www.gobiernodecanarias.org/istac/estadisticas/demografia/poblacion/cifraspadronales/E30245A.html (accessed on 6 March 2023).
25. Martin-Alonso, A.; Abreu-Yanes, E.; Feliu, C.; Mas-Coma, S.; Bargues, M.D.; Valladares, B.; Foronda, P. Intermediate Hosts of Angiostrongylus cantonensis in Tenerife, Spain. *PLoS ONE* **2015**, *10*, e0120686. [CrossRef] [PubMed]
26. Anettová, L.; Izquierdo-Rodriguez, E.; Foronda, P.; Baláž, V.; Novotný, L.; Modrý, D. Endemic lizard *Gallotia galloti* is a paratenic host of invasive *Angiostrongylus cantonensis* in Tenerife, Spain. *Parasitology* **2022**, *149*, 934–939. [CrossRef] [PubMed]
27. López, C.; Clemente, S.; Almeida, C.; Brito, A.; Hernández, M. A genetic approach to the origin of *Millepora* sp. in the eastern Atlantic. *Coral Reefs* **2015**, *34*, 631–638. [CrossRef]
28. Qvarnstrom, Y.; da Silva, A.C.A.; Teem, J.L.; Hollingsworth, R.; Bishop, H.; Graeff-Teixeira, C.; da Silva, A.J. Improved Molecular Detection of *Angiostrongylus cantonensis* in Mollusks and Other Environmental Samples with a Species-Specific Internal Transcribed Spacer 1-Based TaqMan Assay. *Appl. Environ. Microbiol.* **2010**, *76*, 5287–5289. [CrossRef] [PubMed]
29. Izquierdo-Rodriguez, E.; Anettová, L.; Hrazdilová, K.; Foronda, P.; Modrý, D. Range of metastrongylids (superfamily Metastrongyloidea) of veterinary and public health interest present in livers of the endemic lizard *Gallotia galloti* from Tenerife, Canary Islands, Spain. *Parasites Vectors* **2023**, *16*, 81. [CrossRef] [PubMed]
30. Kumar, S.; Stecher, G.; Li, M.; Knyaz, C.; Tamura, K. MEGA X: Molecular Evolutionary Genetics Analysis across Computing Platforms. *Mol. Biol. Evol.* **2018**, *35*, 1547–1549. [CrossRef]
31. Saitou, N.; Nei, M. The neighbor-joining method: A new method for reconstructing phylogenetic trees. *Mol. Biol. Evol.* **1987**, *4*, 406–425.
32. Kimura, M. A simple method for estimating evolutionary rates of base substitutions through comparative studies of nucleotide sequences. *J. Mol. Evol.* **1980**, *16*, 111–120. [CrossRef]
33. Wang, L.-C.; Chao, D.; Chen, E.-R. Experimental infection routes of *Angiostrongylus cantonensis* in mice. *J. Helminthol.* **1991**, *65*, 296–300. [CrossRef]
34. Sugaya, H.; Aoki, M.; Abe, T.; Ishida, K.; Yoshimura, K. Cytokine responses in mice infected with *Angiostrongylus cantonensis*. *Parasitol. Res.* **1996**, *83*, 10–15. [CrossRef]
35. Nogales, M.; Rodríguez Luengo, J.L.; Delgado, G.; Quilis, V.; Trujullo, O. The diet of feral cats (*Felis catus*) on Alegranza Island (north of Lanzarote, Canary Islands). *Folia Zool.* **1992**, *41*, 209–212.
36. Medina, F.M.; Bonnaud, E.; Vidal, E.; Tershy, B.R.; Zavaleta, E.S.; Donlan, C.J.; Keitt, B.S.; Le Corre, M.; Horwath, S.V.; Nogales, M. A global review of the impacts of invasive cats on insular endangered vertebrates. *Glob. Chang. Biol.* **2010**, *17*, 3503–3510. [CrossRef]
37. Bonnaud, E.; Medina, F.M.; Vidal, E.; Nogales, M.; Tershy, B.; Zavaleta, E.; Donlan, C.J.; Keitt, B.; Le Corre, M.; Horwath, S.V. The diet of feral cats on islands: A review and a call for more studies. *Biol. Invasions* **2011**, *13*, 581–603. [CrossRef]
38. Gobierno de Canarias. Banco de Datos de Biodiversidad de Canarias. Available online: http://www.biodiversidadcanarias.es/biota (accessed on 29 March 2023).
39. Kim, J.R.; Hayes, K.A.; Yeung, N.W.; Cowie, R.H. Diverse gastropod hosts of *Angiostrongylus cantonensis*, the rat lungworm, globally and with a focus on the Hawaiian Islands. *PLoS ONE* **2014**, *9*, e94969. [CrossRef] [PubMed]
40. Morgan, E.R.; Jefferies, R.; Krajewski, M.; Ward, P.; Shaw, S.E. Canine pulmonary angiostrongylosis: The influence of climate on parasite distribution. *Parasitol. Int.* **2009**, *58*, 406–410. [CrossRef]
41. Kutz, S.J.; Hoberg, E.P.; Polley, L.; Jenkins, E.J. Global warming is changing the dynamics of Arctic host–parasite systems. *Proc. R. Soc. Lond. Biol.* **2005**, *272*, 2571–2576. [CrossRef] [PubMed]
42. Salb, A.L.; Barkema, H.W.; Elkin, B.T.; Thompson, R.C.A.; Whiteside, D.P.; Black, S.R.; Dubey, J.P.; Kutz, S.J. Dogs as Sources and Sentinels of Parasites in Humans and Wildlife, Northern Canada. *Emerg. Infect. Dis.* **2008**, *14*, 60–63. [CrossRef] [PubMed]
43. Eamsobhana, P. Review paper eosinophilic meningitis caused by *Angiostrongylus cantonensis* a neglected disease with escalating importance. *Trop. Biomed* **2014**, *31*, 569–578. [PubMed]
44. Nguyen, Y.; Rossi, B.; Argy, N.; Baker, C.; Nickel, B.; Marti, H.; Zarrouk, V.; Houzé, S.; Fantin, B.; Lefort, A. Autochthonous case of eosinophilic meningitis caused by *Angiostrongylus cantonensis*, France, 2016. *Emerg. Infect. Dis.* **2017**, *23*, 1045. [CrossRef] [PubMed]
45. Halliday, J.E.; Meredith, A.L.; Knobel, D.L.; Shaw, D.J.; Bronsvoort, B.M.D.C.; Cleaveland, S. A framework for evaluating animals as sentinels for infectious disease surveillance. *J. R. Soc. Interface* **2007**, *4*, 973–984. [CrossRef]
46. Morgan, E.R.; Modry, D.; Paredes-Esquivel, C.; Foronda, P.; Traversa, D. Angiostrongylosis in Animals and Humans in Europe. *Pathogens* **2021**, *10*, 1236. [CrossRef] [PubMed]

animals

Article

High Prevalence of Microsporidia in the North African Hedgehog (*Atelerix algirus*) in the Canary Islands, Spain

Edgar Baz-González [1,2,3], Néstor Abreu-Acosta [2,4] and Pilar Foronda [1,2,*]

1 Department Obstetricia y Ginecología, Pediatría, Medicina Preventiva y Salud Pública, Toxicología, Medicina Legal y Forense y Parasitología, Universidad de La Laguna (ULL), Avda. Astrofísico F. Sánchez s/n, 38203 San Cristóbal de La Laguna, Tenerife, Canary Islands, Spain; ebazgonz@ull.edu.es

2 Instituto Universitario de Enfermedades Tropicales y Salud Pública de Canarias (IUETSPC), Universidad de La Laguna (ULL), Avda. Astrofísico F. Sánchez s/n, 38203 San Cristóbal de La Laguna, Tenerife, Canary Islands, Spain; gerencia@nertalab.es

3 Programa de Doctorado en Ciencias Médicas y Farmacéuticas, Desarrollo y Calidad de Vida, Universidad de La Laguna (ULL), Avda. Astrofísico F. Sánchez s/n, 38203 San Cristóbal de La Laguna, Tenerife, Canary Islands, Spain

4 Nertalab S.L.U., 38001 Santa Cruz de Tenerife, Tenerife, Canary Islands, Spain

* Correspondence: pforonda@ull.edu.es

Simple Summary: It is well known that mammals can harbor various pathogens that can affect humans (known as zoonotic pathogens) including viruses, bacteria, fungi and parasites. Microsporidia are a group of pathogens related to fungi and parasites of several animals that can cause diarrhea or systemic infection in humans. Due to the limited knowledge about microsporidia infection in hedgehogs worldwide, this study aimed to analyze the presence and identity of microsporidia in a group of North African hedgehogs from the Canary Islands (Spain). *Enterocytozoon bieneusi* and *Encephalitozoon cuniculi*, two zoonotic species of microsporidia, were identified. These results suggest that microsporidia species with zoonotic risk circulate in the archipelago.

Abstract: Microsporidia are unicellular eukaryotic obligate intracellular parasites with a wide range of hosts reported worldwide; however, little is known about the epidemiological data on microsporidia infection in animals from the Canary Islands. Since data on microsporidia infection in hedgehog species are scarce, the aim of this study was to analyze the presence and identity of microsporidia in a group of North African hedgehogs (*Atelerix algirus*) using microscopic and molecular methods. From December 2020 to September 2021, a total of 36 fecal samples were collected from naturally deceased hedgehogs from Tenerife and Gran Canaria. All samples showed spore-compatible structures (100%; 36/36) under microscopic analysis, of which 61.1% (22/36) were amplified via the nested-polymerase chain reaction (PCR) targeting the partial sequence of the 16S rRNA gene, the internal transcribed spacer (ITS) region, and the partial sequence of the 5.8S rRNA gene. After Sanger sequencing and ITS analysis, *Enterocytozoon bieneusi* was detected in 47.2% (17/36) of the samples, identifying two novel genotypes (AAE1 and AAE2), followed by the detection of an undetermined species in 8.3% (3/36) and *Encephalitozoon cuniculi* genotype I in 5.6% (2/36) of the samples. This study constitutes the first report of microsporidia species in *Atelerix algirus* worldwide, highlighting the high prevalence of zoonotic species.

Keywords: microsporidia; *Enterocytozoon bieneusi*; *Encephalitozoon cuniculi*; hedgehog; *Atelerix algirus*; zoonotic; Canary Islands

1. Introduction

Microsporidia are unicellular eukaryotic obligate intracellular parasites, spore-forming, and phylogenetically related to the fungi kingdom. Seventeen species have been described as human-pathogenic microsporidia, especially in immunocompromised individuals, of which *Enterocytozoon bieneusi* and the genus *Encephalitozoon* are the most frequent [1].

Enterocytozoon bieneusi, with approximately 500 genotypes described based on the sequence analysis of the internal transcribed spacer (ITS) of the rRNA gene, is clustered within 11 phylogenetic groups. Group 1 and Group 2 are considered zoonotic, while the remaining groups (3–11) are considered host-specific [2]. In *Encephalitozoon cuniculi*, four genotypes (I–IV) based on the number of 5′-GTTT-3′ repeats in the ITS have been described, all of which have confirmed zoonotic potential [3]. The number of genotypes of *Encephalitozoon hellem* depends on the target for genotyping (ITS, the polar tube protein locus, or other intergenic spacers), suggesting high intraspecies variability [4]. However, there are no genetic differences within the ITS for genotyping the *Encephalitozoon intestinalis* isolates [5].

Spore transmission can occur through ingestion of contaminated water and food, inhalation of contaminated aerosols, contact with infected animals (zoonotic transmission) or persons (anthroponotic transmission). Zoonotic transmission has been supported by the identification of the same genotypes in humans and animals [6,7].

Microsporidia have been found in several hosts, including livestock, companion and wildlife animals worldwide, but little is known about microsporidia infection in hedgehogs [8]. To our knowledge, the detection of *E. bieneusi* has only been reported in an undetermined species of hedgehog [9], in the Amur hedgehog (*Erinaceus amurensis*) [10] and more recently in the African pygmy hedgehog (*Atelerix albiventris*) [11].

In the Canary Islands (13°23′–18°8′ W and 27°37′–29°24′ N), the only hedgehog species recorded is the North African hedgehog (*Atelerix algirus*), an introduced species from Northwest Africa. The distribution of this mammal in Spain includes the Iberian Peninsula, the Balearic and Canary Islands, as well as Ceuta and Melilla. The introduction of this species has been suggested as an anthroponotic introduction from Morocco to Fuerteventura in 1892, and is currently present in Fuerteventura, Lanzarote, Gran Canaria, and Tenerife. Nonetheless, isolated specimens are also known from La Gomera, El Hierro, and La Palma [12,13].

Epidemiological data on microsporidia infection in the fauna of this archipelago are scarce, and there are no data on *A. algirus* as a host for these parasites. Therefore, the present study aimed to investigate the prevalence and identification of microsporidia in fecal samples from hedgehogs on the Canary Islands.

2. Materials and Methods

2.1. Ethical Agreement

This study was carried out under the agreement of "Consejería de Transición Ecológica, Lucha contra el Cambio Climático y Planificación Territorial" (Gobierno de Canarias) named "Estudio de patógenos en aves migratorias y en especies exóticas en un escenario de cambio climático", available online in the Official Bulletin of the Canaries (BOC n° 248, 4 December 2020) [14].

2.2. Study Area, Sample Collection, and Preparation

From December 2020 to September 2021, a total of 36 fecal samples were collected by dissecting naturally deceased hedgehogs (n = 33) donated by "La Tahonilla" Wildlife Recovery Center in Tenerife, and found dead individuals (n = 3) collected by technical personnel of "RedEXOS" in Gran Canaria (Figure 1). For each sampled animal, sex and location were recorded whenever possible (Supplementary Material—Table S1).

All fecal samples were placed in tubes containing 2.5% (w/v) aqueous potassium dichromate solution ($K_2Cr_2O_7$) (Merck, Darmstadt, Germany) and stored at 4 °C until further processing.

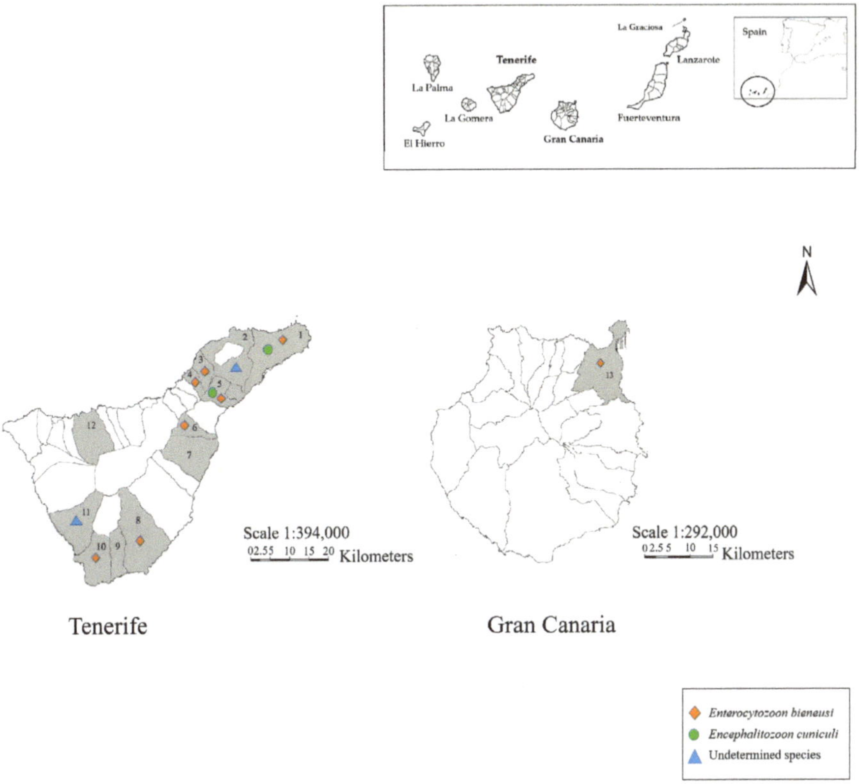

Figure 1. Map of the sampled locations in Tenerife and Gran Canaria. The municipalities are shown in gray—1: Santa Cruz de Tenerife; 2: San Cristóbal de La Laguna; 3: Tacoronte; 4: El Sauzal; 5: El Rosario; 6: Arafo; 7: Güímar; 8: Granadilla de Abona; 9: San Miguel de Abona; 10: Arona; 11: Adeje; 12: Icod de Los Vinos; 13: Las Palmas de Gran Canaria. The symbols indicate samples confirmed via nested-PCR. The original images were taken from Wikimedia Common (https://commons.wikimedia.org/w/index.php?title=File:Mapa_Canarias_municipios.svg&oldid=478721455, accessed on 2 March 2023; https://commons.wikimedia.org/wiki/File:Islas_Canarias_(real_location)_in_Spain.svg, accessed on 2 March 2023) and Gobierno de Canarias (https://www3.gobiernodecanarias.org/medusa/mediateca/ecoescuela/?attachment_id=3333, accessed on 2 March 2023; https://www3.gobiernodecanarias.org/medusa/mediateca/ecoescuela/?attachment_id=3265, accessed on 2 March 2023), in which permission to copy, distribute, or adapt was established. Users: Júlio Reis (https://commons.wikimedia.org/wiki/User:Tintazul, accessed on 2 March 2023), TUBS (https://commons.wikimedia.org/wiki/User:TUBS, accessed on 2 March 2023), GRAFCAN (https://www.grafcan.es/, accessed on 2 March 2023), and IDE Canarias (http://www.idecanarias.es/, accessed on 2 March 2023) (Source: Gobierno de Canarias).

2.3. Staining Method

Fecal samples were stained with Weber's chromotrope stain (chromotrope 2R [Sigma-Aldrich, St. Louis, MO, USA], and Fast Green [Sigma-Aldrich, St. Louis, MO, USA] and phosphotungstic acid [Sigma-Aldrich, St. Louis, MO, USA]) [15] and microscopically screened for microsporidia spores at a magnification of 1000× under a Leica DM750 microscope model ICC50 HD (Leica Microsystems, Heerbrugg, Switzerland). Samples with spore-compatible structures, ovoid and refractile structures stained pink-red, were considered positive.

2.4. DNA Extraction

The total DNA of each fecal sample was extracted using ~500 μL of the sample, previously washed with sterile Phosphate-Buffered Saline (PBS) 1X at room temperature, and centrifuged at 3500 rpm for 15 min to remove the potassium dichromate solution. A commercial FastDNA ® Spin Kit for Soil (MP Biomedicals, Solon, OH, USA) was used following the manufacturer's instructions, and the homogenizer FastPrep-24^{TM} 5G (MP Biomedicals, Solon, OH, USA) was used as the spore disruptor.

2.5. Nested-PCR Amplification

Nested-PCR was carried out in an XP Cycler (Bioer Technology, Hangzhou, China) targeting the partial sequence of the 16S rRNA gene, the complete internal transcribed spacer region (ITS), and the partial sequence of the 5.8S rRNA gene [16].

The amplification reaction of both steps (25 μL) included 0.15 μL of Taq DNA polymerase (5 UI/ μL) (VWR International, Haasrode, Belgium), 2.5 μL of dNTPs mix (200 μM) (Bioline, London, UK), 2.5 μL of 10× key buffer (15 mM Mg^{2+}) (VWR International, Haasrode, Belgium), 1.25 μL of $MgCl_2$ (25 mM) (VWR International, Haasrode, Belgium), 0.1 μL of each primer and 1 μL of DNA template (or 1 μL of primary PCR product). The primers used were MSP-1, MSP-2A, and MSP2B for the first step and MSP-3, MSP-4A, and MSP4B for the second step. The pairs of primers were used to identify *E. bieneusi* (MSP1/MSP2B—MSP3/MSP4B) and *Encephalitozoon* spp. (MSP1/MSP2A—MSP3/MSP4A) with the following conditions in each reaction: initial denaturation at 94 °C for 3 min, 35 cycles of denaturation at 94 °C for 45 s, annealing at 54 °C for 45 s, and extension at 72 °C for 1 min, and a final step at 72 °C for 7 min [17].

Ten microliters of each PCR product were examined via electrophoresis on 1.5% (*w*/*v*) agarose gels (Fisher Bioreagents, Madrid, Spain) stained with REALSAFE Nucleic Acid Staining Solution (20,000×, REAL, Durviz S.L., Valencia, Spain). An amplified DNA product with sizes between 300 and 500 bp (expected size for *Encephalitozoon* spp. and *E. bieneusi*, respectively) was considered positive and was sequenced using secondary primers.

2.6. Sequencing and Phylogenetic Analysis

All nested-PCR positive products were purified using ExoCleanUp FAST (VWR International, Haasrode, Belgium) and sequenced using the Sanger method at the University of La Laguna Sequencing Services (Servicio de Genómica—Servicios Generales de Apoyo a la Investigación de la Universidad de La Laguna, Universidad de La Laguna, Spain).

The obtained sequence chromatograms were analyzed and aligned using the ClustalW program included in MEGA X v10.2.6 (Molecular Evolutionary Genetic Analysis) software (Hachioji, Japan) [18] and compared using the Basic Local Alignment Search Tool (BLAST) in the GenBank database.

Phylogenetic trees were generated using the neighbor-joining method, and genetic distances were calculated using the Kimura 2-parameter model [19,20] with 1000 bootstrap replicates.

Nucleotide sequences were deposited in GenBank under the following accession numbers for *E. bieneusi* (OQ646695–OQ646706; and OQ646730–OQ646734), *E. cuniculi* (OQ646736 and OQ646737) and the undetermined species (OQ646735).

3. Results

3.1. Light Microscopy

Of the 36 hedgehogs, spore-compatible structures were found in 100% (36/36) of the samples stained with Weber's chromotrope stain.

3.2. Molecular Characterization

A total of 22 (61.1%) samples yielded fragments of the expected sizes (300–500 bp). Sanger sequencing revealed the presence of *E. bieneusi* in 47.2% (17/36) and *E. cuniculi* in 5.6% (2/36) of samples. In addition, three samples (8.3%; 3/36) were identified as

undetermined species based on BLAST analysis because of the low homology observed (less than 95%) or because they were not long enough for homology comparison with the reference sequences in MEGA X.

3.2.1. Molecular Characterization of *Enterocytozoon bieneusi*

Genotyping was successful in 94.1% (16/17) of *E. bieneusi*-positive samples. One sequence (5.9%) was not sufficiently long to be genotyped. Two novel genotypes were identified, named AAE1 (*n* = 13) and AAE2 (*n* = 3).

The sequence of the ITS region of genotype AAE1 (242-bp) showed 99.18% homology with genotypes isolated from several mammal species in China: HND-I in snub-nosed monkeys (*Rhinopithecus bieti*) (MK965088.1) and sika deer (*Cervus nippon*) (KX383628.1); and Type IV in raccoon dog (*Nyctereutes procyonoides*) (MN747469.1) and Père David's deer (*Elaphurus davidianus*) (KP057598.1).

The sequence of the ITS region of the genotype AAE2 (243-bp) showed 99.69% homology with various isolates of WildBoar3 (syn. NCF2, NCF3, NCF4) in the silver fox (*Vulpes vulpes*) and arctic fox (*Vulpes lagopus*) from China (MN029056.1), beech marten (*Martes foina*) (MN218601.1) from Poland, European badger (*Meles meles*) (MG458713.1) from Spain, and red fox (*Vulpes vulpes*) from Poland and Spain (MK256483.1 and MG458714.1, respectively).

The ITS region of genotype AAE1 was 242 bp in length, as a result of the deletion of one nucleotide (position 53), differed by one single nucleotide polymorphism (SNP) compared with genotype HND-I, two SNPs compared with genotype EA1, one SNP and two nucleotide insertions compared with genotype EA2, two SNPs compared with genotype EA3, one SNP compared with EA4, and two SNPs compared with genotype S9. The positions of the SNPs and the insertions are listed in Table 1.

Table 1. Sequence differences in the internal transcribed spacer region of the rRNA gene of the novel genotype AAE1 compared to the closest matched sequences.

Genotype (Host)	Nucleotide Position (5′→ 3′) [1]							
	31	32	51	52	53	86	131	155
AAE1 (*Atelerix algirus*)	A	T	G	T	-	G	G	A
HND-I (*Rhinopithecus bieti*)	G	T	G	T	A	G	G	A
EA1 (*Erinaceus amurensis*)	G	C	G	T	A	G	G	A
EA2 (*Erinaceus amurensis*)	G	T	-	-	A	G	G	A
EA3 (*Erinaceus amurensis*)	G	T	G	T	A	A	G	A
EA4 (*Erinaceus amurensis*)	G	T	G	T	A	G	G	G
S9 (*Vulpes vulpes*)	G	T	G	T	A	G	A	A

[1] Nucleotide positions in the internal transcribed spacer region (ITS) of *Enterocytozoon bieneusi* (~243-bp). Hyphen indicates a deletion in this position.

The ITS region of genotype AAE2, 243 bp in length, showed only one SNP at position 104 (G→ A) compared to genotype WildBoar3.

The novel genotypes clustered within Group 1 based on phylogenetic analysis (Figure 2). Genotype AAE1 fell into an independent clade close to two clades, one of which was formed by the genotypes isolated from the Amur hedgehog (*E. amurensis*) in China (EA1–EA4) (bootstrap value of 100%). Genotype AAE2 clustered in a clade with the genotype WildBoar3 (syn. NCF2, NCF3, NCF4), which have been previously detected in carnivores in mainland Spain.

3.2.2. Molecular Characterization of *Encephalitozoon cuniculi*

Two of the thirty-six (5.6%) fecal samples were positive for *E. cuniculi*. Both sequences showed >99% homology with *E. cuniculi* sequences deposited in GenBank (AB713183.1, L13332.1, and OP555067.1). Phylogenetic analysis confirmed the identity of the isolates as *E. cuniculi* (bootstrap 100%) (Figure 3) and were identified as genotype I based on ITS sequence analysis.

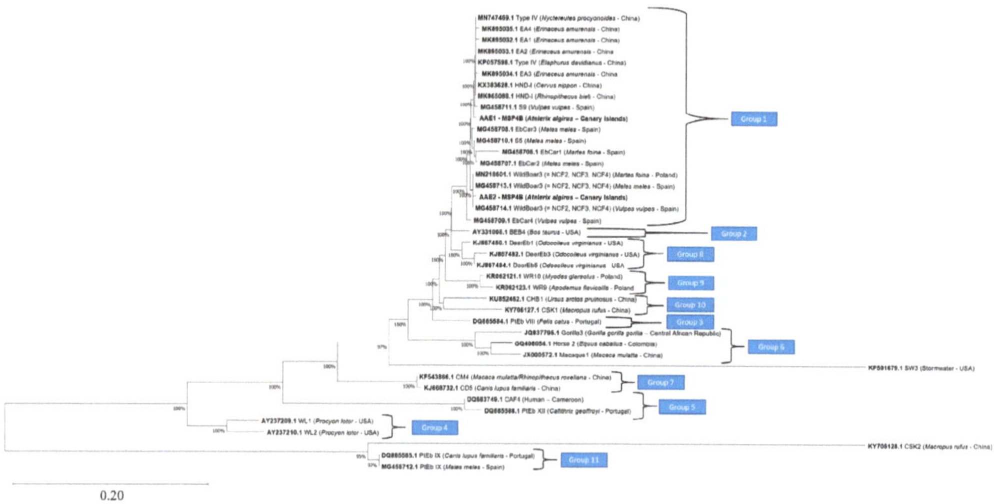

Figure 2. Phylogenetic relationships between the sequence of the internal transcribed spacer region of the rRNA gene of *Enterocytozoon bieneusi* obtained in this study and the sequences of known genotypes deposited in GenBank. The tree was constructed using the neighbor-joining method based on the genetic distance calculated using the Kimura 2-parameter model. Representative sequences from each phylogenetic Group of *E. bieneusi* genotypes (Groups 1–11) were used. Accession numbers are shown in bold, and information regarding the host species and origin are shown in parentheses. There were a total of 239 positions in the final dataset.

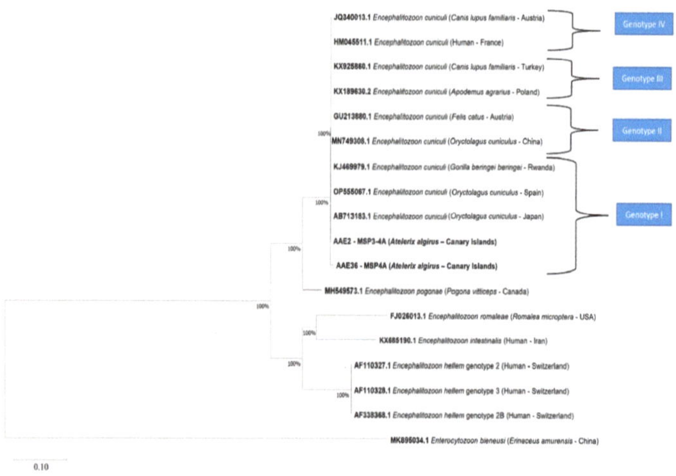

Figure 3. Phylogenetic relationships between partial sequences of the 16S rRNA gene, the complete internal transcribed spacer region (ITS), and the partial sequence of the 5.8S rRNA gene of *Encephalitozoon cuniculi* obtained in this study and known genotype sequences deposited in GenBank. The tree was constructed using the neighbor-joining method based on the genetic distance calculated using the Kimura 2-parameter model. Representative sequences for each *E. cuniculi* genotype (I–IV) were used. Accession numbers are shown in bold, and information concerning the host species and origin are shown in parentheses. *Enterocytozoon bieneusi* (MK895034.1) was used as an outgroup. There were a total of 211 positions in the final dataset.

3.2.3. Geographical Distribution

Of the 16 genotyped *E. bieneusi* isolates, the most frequently detected was genotype AAE1 (81.25%; 13/16). It showed a wide distribution, being detected in seven municipalities in Tenerife and in the sampled municipality in Gran Canaria. In contrast, the genotype AAE2 (18.75%; 3/16) was detected only in two northern municipalities of Tenerife Island, El Sauzal, and Tacoronte.

Encephalitozoon cuniculi-positive hedgehogs were found in the northeastern zone of Tenerife, specifically in Santa Cruz de Tenerife and El Rosario.

Undetermined species were detected in Adeje and San Cristóbal de La Laguna; no other species were detected at these locations (Figure 1, Table 2).

Table 2. Geographical distribution of microsporidia species and genotypes identified in *Atelerix algirus* on the Canary Islands.

Location	Sample Size (*n*)	Nested-PCR Confirmed Samples (*n*)	*E. bieneusi* Genotypes (*n*)	*E. cuniculi* Genotypes (*n*)	Undetermined Species
Adeje	1	1	-	-	1
Arafo	1	-	-	-	-
Arona	7	6	AAE1 (5) Undetermined genotype (1) *	-	-
El Rosario	3	2	AAE1 (1)	I (1)	-
El Sauzal	1	1	AAE2 (1)	-	-
Granadilla de Abona	3	2	AAE1 (2)	-	-
Güímar	1	-	-	-	-
Icod de Los Vinos	1	-	-	-	-
Las Palmas de Gran Canaria	3	3	AAE1 (3)	-	-
San Cristóbal de La Laguna	7	2	-	-	2
San Miguel de Abona	1	-	-	-	-
Santa Cruz de Tenerife	5	3	AAE1 (2)	I (1)	-
Tacoronte	2	2	AAE2 (2)	-	-
TOTAL	36	22	17	2	3

* The sample was not successfully genotyped.

4. Discussion

The present study constitutes the first report of microsporidia in the North African hedgehog (*A. algirus*), highlighting the presence of human-pathogenic microsporidia, *E. bieneusi* and *E. cuniculi*.

In Spain, *E. bieneusi* has been reported as the most frequent etiological agent of intestinal microsporidiosis in patients with human immunodeficiency virus (HIV) [21–29], as well as in non-HIV patients [30–36], and in sporadic cases of extraintestinal microsporidiosis [21,22,28,30,34,35,37]. However, cases of *Encephalitozoon* spp. have rarely been identified in Spain [29,32,38–42].

Data on microsporidia infection in patients from the Canary Islands are scarce, with only two reports of *E. bieneusi* in immunocompetent patients from Tenerife [35] and transplant recipients from Gran Canaria (genotype D) [36].

Regarding the zoonotic role of microsporidia, several animal hosts have been described as reservoirs of this group of parasites. To date, *E. bieneusi* is the most common species in Spain and has been detected in pet dogs [43–46], wild animals (lagomorphs, rodents, carnivores, and ungulates) [43,47–52], animals in urban environments (pigeons and cats) [46,53,54], farm animals (goats, rabbits, pigs, ostriches, cattle, and deer) [43–45,51,52,55,56] and animals in zoos (chimpanzees) [57]. In studies wherein genotyping was performed, most of the detected genotypes clustered within Group 1, suggesting zoonotic potential. *Encephalitozoon intestinalis* is the second most frequent species in animals. It has been reported in domestic cats [45], wildlife rabbits and hares [47,58], pigeons from parks [53], and farmed pigs and ostriches [45], whereas there are only a few reports of *E. hellem* [53,59] and *E. cuniculi* [50,60,61].

The prevalence obtained using the microscopic method in this study (100%; 36/36) differed from that obtained using nested-PCR (61.1%; 22/36), as reported in other studies [46,50]. This difference can be explained by the spontaneous extrusion of spores or low parasitic load, as suggested in the studies conducted by Izquierdo et al. [50] and Haro et al. [53], respectively.

The overall prevalence obtained in animals using the molecular methods in studies conducted in Spain with similar sample sizes ranged from 43.8% (14/32) in domestic dogs in Madrid [45], to 55.6% (15/27) in farmed pigs in Extremadura and Castile and León, [45] and 65.4% (17/26) in Iberian lynx in Andalusia [50], with *E. bieneusi* being most commonly detected in fecal samples in the latter studies. However, *E. bieneusi* was detected in 7 of the 14 PCR-positive dog samples and in 7 of the 15 PCR-positive pig samples compared to the 17 of the 22 nested-PCR-positive samples detected in this study.

In the case of the Iberian lynx, the results were in agreement with the positive samples obtained in this study for *E. bieneusi* (76.5%; 13/17 in lynxes vs. 77.3%; 17/22 in hedgehogs) and *E. cuniculi* (11.8%; 2/17 in lynxes vs. 9.1%; 2/22 in hedgehogs).

Considering the host species, the 47.2% (17/36) prevalence of *E. bieneusi* reported in *A. algirus* fecal samples is higher than that reported in a study conducted in China, which reported a prevalence of 9.8% (4/41) in *E. amurensis* intestine samples [10], but lower than the 70.0% (266/380) reported in fecal samples from farmed and pet *A. albiventris,* also conducted in China. The highest genetic diversity was recorded in *A. albiventris* with one known genotype, SCR05 (88.3%; 235/266) and 10 novel genotypes, GDH01 (3.4%; 9/266), GDH02 (0.8%; 3/266), and GDH03– GDH10 (one sample each) [11], followed by *E. amurensis* with four novel genotypes, EA1, EA2, EA3, and EA4 (one sample each) [10]. The population of *A. algirus* on the Canary Islands showed low genetic diversity and two novel genotypes, AAE1 and AAE2.

To the best of our knowledge, this is the first study to detect *E. cuniculi* in hedgehogs. The prevalence obtained (5.6%; 2/36) was similar to that recently reported in fecal samples from European rabbits in Tenerife (4.0%; 2/50) [61]. Genotype I of *E. cuniculi* has been the only genotype detected in animal hosts in Spain to date [60,61], and the species is less frequent in this country, with a few cases in humans [40–42]. However, molecular detection in water sources, in addition to serological analysis of domestic and wild animals, demonstrates the presence of this parasite in the environment [50,61,62].

Considering that the diet of hedgehogs is mainly based on Coleoptera [63] and numerous species of microsporidia have been described as insect parasites [64], the undetermined species detected in fecal samples of hedgehogs are suspected to be microsporidia species infecting invertebrate hosts. Other studies have also detected undetermined species using molecular methods [45,61,62].

The remaining human-pathogenic *Encephalitozoon* species, *E. intestinalis* and *E. hellem* were not detected in this study. A low prevalence of *E. intestinalis* has been reported in animals in Spain [45,47,53,58] and *E. hellem* in mammals worldwide [65].

Considering the high prevalence of *E. bieneusi* genotypes with zoonotic potential, veterinary control measures should be implemented to detect this pathogen, given that hedgehogs have been kept as pets on the Canary Islands [66] and could pose a risk to children who are most susceptible to microsporidiosis [67]. Despite the limited number of *E. cuniculi* cases detected in this study, the zoonotic risk should not be underestimated because symptomatic cases have been documented in humans [9].

5. Conclusions

This study constitutes the first report of microsporidia in fecal samples of the North African hedgehog (*A. algirus*). The overall prevalence of nested-PCR-confirmed samples was 61.1% (22/36), with *E. bieneusi* being the most common species, followed by the undetermined species and *E. cuniculi*. Two novel genotypes of *E. bieneusi* were identified, named AAE1 and AAE2, both clustered within Group 1, and the *E. cuniculi* isolates were identified as genotype I.

The results obtained in this study provide new data on the epidemiology of microsporidia on the Canary Islands (Spain), suggesting that zoonotic genotypes of human-pathogenic microsporidia circulate in the fauna of the islands, posing a risk to public and veterinary health.

Supplementary Materials: The following supporting information can be downloaded at: https://www.mdpi.com/article/10.3390/ani13111756/s1, Table S1: Microsporidia detected in hedgehog (*Atelerix algirus*) fecal samples and location in the Canary Islands (Spain).

Author Contributions: Conceptualization, P.F. and N.A.-A.; methodology, E.B.-G.; formal analysis, E.B.-G.; investigation, E.B.-G.; resources, P.F.; data curation, E.B.-G., P.F. and N.A.-A.; writing—original draft preparation, E.B.-G. and P.F.; writing—review and editing, N.A.-A.; supervision, P.F.; project administration, P.F.; funding acquisition, P.F. All authors have read and agreed to the published version of the manuscript.

Funding: This research was funded by ProID2021010013 Consejería de Economía, Industria, Comercio y Conocimiento (Gobierno de Canarias) and Fondos FEDER-FSE Canarias 2014–2020, "Programa de Apoyo a la Investigación María del Carmen Betencourt y Molina"; and by "Consejería de Transición Ecológica, Lucha contra el Cambio Climático y Planificación Territorial (Gobierno de Canarias)—Universidad de La Laguna agreement". E.B.-G. was granted an FPI predoctoral scholarship by "Agencia Canaria de Investigación, Innovación y Sociedad de la Información de la Consejería de Economía, Conocimiento y Empleo" and "Fondo Social Europeo (FSE) Programa Operativo Integrado de Canarias 2014–2020, Eje 3 Tema Prioritario 74 (85%)" (TESIS2021010056).

Institutional Review Board Statement: We consider that the Ethical Declaration is not necessary since in this study only dead or naturally deceased hedgehogs were used.

Informed Consent Statement: The head of the Wildlife Recovery Center has signed the consent form.

Data Availability Statement: Not applicable.

Acknowledgments: We wish to thank the staff of "La Tahonilla" Wildlife Recovery Center and "RedEXOS" for providing the animals.

Conflicts of Interest: The funders had no role in the design of the study; in the collection, analyses, or interpretation of data; in the writing of the manuscript; or in the decision to publish the results.

References

1. Han, B.; Pan, G.; Weiss, L.M. Microsporidiosis in Humans. *Clin. Microbiol. Rev.* **2021**, *34*, e0001020. [CrossRef] [PubMed]
2. Li, W.; Feng, Y.; Santin, M. Host Specificity of *Enterocytozoon bieneusi* and Public Health Implications. *Trends Parasitol.* **2019**, *35*, 436–451. [CrossRef] [PubMed]
3. Santaniello, A.; Cimmino, I.; Dipineto, L.; Agognon, A.L.; Beguinot, F.; Formisano, P.; Fioretti, A.; Menna, L.C.; Oriente, F. Zoonotic Risk of *Encephalitozoon cuniculi* in Animal-Assisted Interventions: Laboratory Strategies for the Diagnosis of Infections in Humans and Animals. *Int. J. Environ. Res. Public Health* **2021**, *18*, 9333. [CrossRef]
4. Haro, M.; Del Águila, C.; Fenoy, S.; Henriques-Gil, N. Intraspecies genotype variability of the microsporidian parasite *Encephalitozoon hellem*. *J. Clin. Microbiol.* **2003**, *41*, 4166–4171. [CrossRef] [PubMed]
5. Graczyk, T.K.; Bosco-Nizeyi, J.; da Silva, A.J.; Moura, I.N.; Pieniazek, N.J.; Cranfield, M.R.; Lindquist, H.D. A single genotype of *Encephalitozoon intestinalis* infects free-ranging gorillas and people sharing their habitats in Uganda. *Parasitol. Res.* **2002**, *88*, 926–931. [CrossRef] [PubMed]
6. Didier, E.S.; Stovall, M.E.; Green, L.C.; Brindley, P.J.; Sestak, K.; Didier, P.J. Epidemiology of microsporidiosis: Sources and modes of transmission. *Vet. Parasitol.* **2004**, *126*, 145–166. [CrossRef] [PubMed]
7. Stentiford, G.D.; Becnel, J.J.; Weiss, L.M.; Keeling, P.J.; Didier, E.S.; Williams, B.A.P.; Bjornson, S.; Kent, M.L.; Freeman, M.A.; Brown, M.J.F.; et al. Microsporidia—Emergent Pathogens in the Global Food Chain. *Trends Parasitol.* **2016**, *32*, 336–348. [CrossRef]

8. Koehler, A.V.; Zhang, Y.; Gasser, R.B. A Perspective on the Molecular Identification, Classification, and Epidemiology of *Enterocytozoon bieneusi* of Animals. *Exp. Suppl.* **2022**, *114*, 389–415.
9. Mathis, A.; Weber, R.; Deplazes, P. Zoonotic potential of the microsporidia. *Clin. Microbiol. Rev.* **2005**, *18*, 423–425. [CrossRef]
10. Gong, X.Q.; Xiao, X.; Gu, X.L.; Han, H.J.; Yu, X.J. Detection of Intestinal Parasites and Genetic Characterization of *Enterocytozoon bieneusi* in Hedgehogs from China. *Vector Borne Zoonotic Dis.* **2021**, *21*, 63–66. [CrossRef]
11. Meng, X.; Chu, W.; Tang, Y.; Wang, W.; Chen, Y.; Li, N.; Feng, Y.; Xiao, L.; Guo, Y. High zoonotic potential and heavy environmental burden of *Cryptosporidium* spp. and *Enterocytozoon bieneusi* in farmed and pet African pygmy hedgehogs (*Atelerix albiventris*). *One Health* **2023**, *16*, 100532. [CrossRef]
12. Banco de Datos de Biodiversidad de Canarias (EXOS). Available online: https://www.biodiversidadcanarias.es/exos/especie/V00149 (accessed on 22 March 2023).
13. Khaldi, M.; Ribas, A.; Barech, G.; Hugot, J.P.; Benyettou, M.; Albane, L.; Arrizabalaga, A.; Nicolas, V. Molecular evidence supports recent anthropogenic introduction of the Algerian hedgehog *Atelerix algirus* in Spain, Balearic and Canary islands from North Africa. *Mammalia* **2016**, *80*, 313–320. [CrossRef]
14. Boletín Oficial de Canarias (BOC). Available online: http://www.gobiernodecanarias.org/boc/2020/248/016.html (accessed on 22 March 2023).
15. Weber, R.; Bryan, R.T.; Owen, R.L.; Wilcox, C.M.; Gorelkin, L.; Visvesvara, G.S. Improved light-microscopical detection of microsporidia spores in stool and duodenal aspirates. The Enteric Opportunistic Infections Working Group. *N. Engl. J. Med.* **1992**, *326*, 161–166. [CrossRef] [PubMed]
16. Katzwinkel-Wladarsch, S.; Lieb, M.; Helse, W.; Löscher, T.; Rinder, H. Direct amplification and species determination of microsporidian DNA from stool specimens. *Trop. Med. Int. Health* **1996**, *1*, 373–378. [CrossRef]
17. Sak, B.; Kváč, M.; Květoňová, D.; Albrecht, T.; Piálek, J. The first report on natural *Enterocytozoon bieneusi* and *Encephalitozoon* spp. infections in wild East-European House Mice (*Mus musculus musculus*) and West-European House Mice (*M. m. domesticus*) in a hybrid zone across the Czech Republic-Germany border. *Vet. Parasitol.* **2011**, *178*, 246–250. [CrossRef] [PubMed]
18. Kumar, S.; Stecher, G.; Li, M.; Knyaz, C.; Tamura, K. MEGA X: Molecular Evolutionary Genetics Analysis across Computing Platforms. *Mol. Biol. Evol.* **2018**, *35*, 1547–1549. [CrossRef]
19. Saitou, N.; Nei, M. The neighbor-joining method: A new method for reconstructing phylogenetic trees. *Mol. Biol. Evol.* **1987**, *4*, 406–425.
20. Kimura, M. A simple method for estimating evolutionary rates of base substitutions through comparative studies of nucleotide sequences. *J. Mol. Evol.* **1980**, *16*, 111–120. [CrossRef]
21. Candia, A.; Lopez-Velez, R.; Perez-Elias, M.J.; Guerrero, A.; Moreira, V.; Redondo, C. *Enterocytozoon bieneusi* Infection in an AIDS patient in Spain. *Res. Rev. Parasitol.* **1994**, *54*, 269–271.
22. Miró, O.; Moreno, A.; Valls, M.E.; Miró, J.M.; Piqué, J.M.; Bordas, J.M.; Moreno-Martínez, A.; Bombí, J.A.; Gatell, J.M. Intestinal microsporidiosis in patients with AIDS. Report of three cases. *Med. Clin. (Barc)* **1995**, *104*, 96–99.
23. Subirats, M.; González-Castelao, G.; Aguilera, O.; Moody, A.; Visvesvara, G.; Verdejo, J.; Baquero, M.; del Águila, C. Diagnosis of 4 cases of intestinal microsporidiosis in AIDS patients. *Enferm. Infecc. Microbiol. Clin.* **1996**, *14*, 533–537. [PubMed]
24. Da Silva, A.J.; Bornay-Llinares, F.J.; del Aguila-de la Puente, C.; Moura, H.; Peralta, J.M.; Sobottka, I.; Schwartz, D.A.; Visvesvara, G.S.; Slemenda, S.B.; Pieniazek, N.J. Diagnosis of *Enterocytozoon bieneusi* (Microsporidia) Infections by Polymerase Chain Reaction in Stool Samples Using Primers Based on the Region Coding for Small-Subunit Ribosomal RNA. *Arch. Pathol. Lab. Med.* **1997**, *121*, 874–879. [PubMed]
25. Moreno-Camacho, A.; Moreno-Martínez, A.; Valls, M.E.; Bordas, J.M.; Piqué, J.M.; Bombí, J.A.; Miró, J.M.; Mallolas, J.; Gatell, J.M.; Soriano, E. Chronic enteropathy of unknown etiology in patients with AIDS. An analysis of 40 cases. *Med. Clin. (Barc)* **1997**, *109*, 452–456. [PubMed]
26. Del Águila, C.; Soriano, V.; Navajas, R.; Subirats, M.; Fenoy, S.; Valencia, E.; Baquero, M.; Pieniazek, N.J. Species identification of intestinal microsporidiosis in HIV-positive patients using the polymerase chain reaction. *Enferm. Infecc. Microbiol. Clin.* **1997**, *15*, 456–461.
27. Del Águila, C.; Navajas, R.; Gurbindo, D.; Ramos, J.T.; Mellado, M.J.; Fenoy, S.; Muñoz-Fernandez, M.A.; Subirats, M.; Ruiz, J.; Pieniazek, N.J. Microsporidiosis in HIV-positive children in Madrid (Spain). *J. Eukaryot. Microbiol.* **1997**, *44*, 84S–85S. [CrossRef]
28. Lores, B.; Arias, C.; Fenoy, S.; Iglesias, I.; Ocampo, A.; Miralles, C.; Garcia-Estevez, J.M.; del Aguila, C. *Enterocytozoon bieneusi*: A common opportunistic parasite in lungs of HIV-positive patients? *J. Eukaryot. Microbiol.* **1999**, *46*, 6S–7S.
29. Chozas, M.; Dashti, A.; Prieto-Pérez, L.; Pérez-Tanoira, R.; Cobo, E.; Bailo, B.; del Palacio, M.; Hernández-Castro, C.; González-Barrio, D.; Carmena, D.; et al. *Enterocytozoon bieneusi* and *Encephalitozoon intestinalis* (Microsporidia) in HIV-positive patients in central Spain. *Med. Mycol.* **2023**, *61*, myad039. [CrossRef]
30. Gainzarain, J.C.; Canut, A.; Lozano, M.; Labora, A.; Carreras, F.; Fenoy, S.; Navajas, R.; Pieniazek, N.J.; da Silva, A.J.; del Águila, C. Detection of *Enterocytozoon bieneusi* in two human immunodeficiency virus-negative patients with chronic diarrhea by polymerase chain reaction in duodenal biopsy specimens and review. *Clin. Infect. Dis.* **1998**, *27*, 394–398. [CrossRef]
31. López-Vélez, R.; Turrientes, M.C.; Garrón, C.; Montilla, P.; Navajas, R.; Fenoy, S.; del Águila, C. Microsporidiosis in travelers with diarrhea from the tropics. *J. Travel Med.* **1999**, *6*, 223–227. [CrossRef]
32. Lores, B.; Arias, C.; López-Miragaya, I.; Torres, J.; Fenoy, S.; del Águila, C. Molecular Diagnosis of Intestinal Microsporidiosis in pediatric patients from Vigo (NW Spain). *Res. Rev. Parasitol.* **2001**, *61*, 43–49.

33. Lores, B.; López-Miragaya, I.; Arias, C.; Fenoy, S.; Torres, J.; del Águila, C. Intestinal microsporidiosis due to *Enterocytozoon bieneusi* in elderly human immunodeficiency virus-negative patients from Vigo, Spain. *Clin. Infect. Dis.* **2002**, *34*, 918–921. [CrossRef]
34. Lores, B.; del Águila, C.; López-Miragaya, I.; Torres, J.; Arias, C. Detection of *Enterocytozoon bieneusi* by PCR in Immunocompetent Spanish patients with diarrhea and pneumonia. *Res. Rev. Parasitol.* **2002**, *62*, 43–49.
35. Abreu-Acosta, N.; Lorenzo-Morales, J.; Leal-Guio, Y.; Coronado-Alvarez, N.; Foronda, P.; Alcoba-Florez, J.; Izquierdo, F.; Batista-Díaz, N.; Del Águila, C.; Valladares, B. *Enterocytozoon bieneusi* (microsporidia) in clinical samples from immunocompetent individuals in Tenerife, Canary Islands, Spain. *Trans. R. Soc. Trop. Med. Hyg.* **2005**, *99*, 848–855. [CrossRef] [PubMed]
36. Galván, A.L.; Sánchez, A.M.; Valentín, M.A.; Henriques-Gil, N.; Izquierdo, F.; Fenoy, S.; del Águila, C. First cases of microsporidiosis in transplant recipients in Spain and review of the literature. *J. Clin. Microbiol.* **2011**, *49*, 1301–1306. [CrossRef] [PubMed]
37. Del Aguila, C.; Lopez-Velez, R.; Fenoy, S.; Turrientes, C.; Cobo, J.; Navajas, R.; Visvesvara, G.S.; Croppo, G.P.; Da Silva, A.J.; Pieniazek, N.J. Identification of *Enterocytozoon bieneusi* spores in respiratory samples from and AIDS patient with a 2-year history of intestinal microsporidiosis. *J. Clin. Microbiol.* **1997**, *35*, 1862–1866. [CrossRef]
38. Peman, J.; Bornay-Llinares, F.J.; Acosta, B.; Lopez-Aldeguer, J.; Meseguer, I.; Figueras, M.J.; Hernandez, A.; Peset, V.; Gobernado, M.; Visvesvara, G.S. First report of a case of *Encephalitozoon* sp. microsporidiosis in a Spanish AIDS patient. *Res. Rev. Parasitol.* **1997**, *57*, 131–134.
39. Bornay-Llinares, F.J.; Acosta, B.; Moura, J.H.; Schwartz, D.A.; da Silva, A.J.; Visvesvara, G.S.; Figueras, M.J.; Gobernado, M.; Pieniazek, N.J. Mantenimiento en cultivo y caracterización de un microsporidio (*Encephalitozoon hellem*) aislado en un paciente con Sida y neumonía. *Parasitología* **2000**, *24*, 69–70. [CrossRef]
40. Del Aguila, C.; Moura, H.; Fenoy, S.; Navajas, R.; Lopez-Velez, R.; Li, L.; Xiao, L.; Leitch, G.J.; da Silva, A.; Pieniazek, N.J.; et al. In vitro culture, ultrastructure, antigenic, and molecular characterization of *Encephalitozoon cuniculi* from urine and sputum samples from a Spanish patient with AIDS. *J. Clin. Microbiol.* **2001**, *39*, 1105–1108. [CrossRef]
41. Andreu-Ballester, J.C.; Garcia-Ballesteros, C.; Amigo, V.; Ballester, F.; Gil-Borrás, R.; Catalán-Serra, I.; Magnet, A.; Fenoy, S.; del Aguila, C.; Ferrando-Marco, J.; et al. Microsporidia and its relation to Crohn's disease. A retrospective study. *PLoS ONE* **2013**, *8*, e62107. [CrossRef]
42. Redondo, F.; Hurtado-Marcos, C.; Izquierdo, F.; Cuéllar, C.; Fenoy, S.; Sáez, Y.; Magnet, A.; Galindo-Regal, L.; Uribe, N.; López-Bañeres, M.; et al. Latent Microsporidia Infection Prevalence as a Risk Factor in Colon Cancer Patients. *Cancers* **2022**, *14*, 5342. [CrossRef]
43. Del Aguila, C.; Izquierdo, F.; Navajas, R.; Pieniazek, N.J.; Miró, G.; Alonso, A.I.; da Silva, A.J.; Fenoy, S. *Enterocytozoon bieneusi* in animals: Rabbits and dogs as new hosts. *J. Eukaryot. Microbiol.* **1999**, *46*, 8S–9S. [PubMed]
44. Lores, B.; del Aguila, C.; Arias, C. *Enterocytozoon bieneusi* (microsporidia) in faecal samples from domestic animals from Galicia, Spain. *Mem. Inst. Oswaldo Cruz* **2002**, *97*, 941–945. [CrossRef] [PubMed]
45. Galván-Díaz, A.L.; Magnet, A.; Fenoy, S.; Henriques-Gil, N.; Haro, M.; Ponce-Gordo, F.; Millán, J.; Miró, G.; del Águila, C.; Izquierdo, F. Microsporidia detection and genotyping study of human pathogenic *E. bieneusi* in animals from Spain. *PLoS ONE* **2014**, *9*, e92289. [CrossRef] [PubMed]
46. Dashti, A.; Santín, M.; Cano, L.; de Lucio, A.; Bailo, B.; Hernández-de Mingo, M.; Köster, P.C.; Fernández-Basterra, J.A.; Aramburu-Aguirre, J.; López-Molina, N.; et al. Occurrence and genetic diversity of *Enterocytozoon bieneusi* (Microsporidia) in owned and sheltered dogs and cats in Northern Spain. *Parasitol. Res.* **2019**, *118*, 2979–2987. [CrossRef]
47. Martínez-Padilla, A.; Caballero-Gómez, J.; Magnet, A.; Gómez-Guillamón, F.; Izquierdo, F.; Camacho-Sillero, L.; Jiménez-Ruiz, S.; del Águila, C.; García-Bocanegra, I. Zoonotic Microsporidia in Wild Lagomorphs in Southern Spain. *Animals* **2020**, *10*, 2218. [CrossRef]
48. Vioque, F.; Dashti, A.; Santín, M.; Ruiz-Fons, F.; Köster, P.C.; Hernández-Castro, C.; García, J.T.; Bailo, B.; Ortega, S.; Olea, P.P.; et al. Wild micromammal host spectrum of zoonotic eukaryotic parasites in Spain. Occurrence and genetic characterisation. *Transbound. Emerg. Dis.* **2022**, *69*, e2926–e2942. [CrossRef]
49. Santín, M.; Calero-Bernal, R.; Carmena, D.; Mateo, M.; Balseiro, A.; Barral, M.; Lima-Barbero, J.F.; Habela, M.A. Molecular Characterization of *Enterocytozoon bieneusi* in Wild Carnivores in Spain. *J. Eukaryot. Microbiol.* **2018**, *65*, 468–474. [CrossRef]
50. Izquierdo, F.; Ollero, D.; Magnet, A.; Galván-Díaz, A.L.; Llorens, S.; Vaccaro, L.; Hurtado-Marcos, C.; Valdivieso, E.; Miró, G.; Hernández, L.; et al. Microsporidia as a Potential Threat to the Iberian Lynx (*Lynx pardinus*). *Animals* **2022**, *12*, 2507. [CrossRef]
51. Dashti, A.; Rivero-Juarez, A.; Santín, M.; López-López, P.; Caballero-Gómez, J.; Frías-Casas, M.; Köster, P.C.; Bailo, B.; Calero-Bernal, R.; Briz, V.; et al. *Enterocytozoon bieneusi* (Microsporidia): Identification of novel genotypes and evidence of transmission between sympatric wild boars (*Sus scrofa ferus*) and Iberian pigs (*Sus scrofa domesticus*) in Southern Spain. *Transbound. Emerg. Dis.* **2020**, *67*, 2869–2880. [CrossRef]
52. Dashti, A.; Santín, M.; Köster, P.C.; Bailo, B.; Ortega, S.; Imaña, E.; Habela, M.A.; Rivero-Juarez, A.; Vicente, J.; WE&H Group; et al. Zoonotic *Enterocytozoon bieneusi* genotypes in free-ranging and farmed wild ungulates in Spain. *Med. Mycol.* **2022**, *60*, myac070. [CrossRef]
53. Haro, M.; Izquierdo, F.; Henriques-Gil, N.; Andrés, I.; Alonso, F.; Fenoy, S.; del Aguila, C. First detection and genotyping of human-associated microsporidia in pigeons from urban parks. *Appl. Environ. Microbiol.* **2005**, *71*, 3153–3157. [CrossRef] [PubMed]
54. Haro, M.; Henriques-Gil, N.; Fenoy, S.; Izquierdo, F.; Alonso, F.; Del Aguila, C. Detection and genotyping of *Enterocytozoon bieneusi* in pigeons. *J. Eukaryot. Microbiol.* **2006**, *53*, S58–S60. [CrossRef]

55. Dashti, A.; Rivero-Juárez, A.; Santín, M.; George, N.S.; Köster, P.C.; López-López, P.; Risalde, M.A.; García-Bocanegra, I.; Gómez-Villamandos, J.C.; Caballero-Gómez, J.; et al. Diarrhoea-causing enteric protest species in intensively and extensively raised pigs (*Sus scrofa domesticus*) in Southern Spain. Part I: Prevalence and genetic diversity. *Transbound. Emerg. Dis.* **2022**, *69*, e1051–e1064. [CrossRef] [PubMed]
56. Abarca, N.; Santín, M.; Ortega, S.; Maloney, J.G.; George, N.S.; Molokin, A.; Cardona, G.A.; Dashti, A.; Köster, P.C.; Bailo, B.; et al. Molecular Detection and Characterization of *Blastocystis* sp. and *Enterocytozoon bieneusi* in Cattle in Northern Spain. *Vet. Sci.* **2021**, *8*, 191. [CrossRef] [PubMed]
57. Köster, P.C.; Martínez-Nevado, E.; González, A.; Abelló-Poveda, M.T.; Fernández-Bellon, H.; de la Riva-Fraga, M.; Marquet, B.; Guéry, J.P.; Knauf-Witzens, T.; Weigold, A.; et al. Intestinal Protists in Captive Non-human Primates and Their Handlers in Six European Zoological Gardens. Molecular Evidence of Zoonotic Transmission. *Front. Vet. Sci.* **2022**, *8*, 819887. [CrossRef]
58. Rego, L.; Castro-Scholten, S.; Cano, C.; Jiménez-Martín, D.; Köster, P.C.; Caballero-Gómez, J.; Bailo, B.; Dashti, A.; Hernández-Castro, C.; Cano-Terriza, D.; et al. Iberian wild leporidae as hosts of zoonotic enteroparasites in Mediterranean ecosystems of Southern Spain. *Zoonoses Public Health* **2022**, *70*, 223–237. [CrossRef]
59. Rosell, J.; Máinez, M.; Didier, E.S.; Bowers, L.C.; Marco, A.; Juan-Sallés, C. *Encephalitozoon hellem* infection in aviary passerine and psittacine birds in Spain. *Vet. Parasitol.* **2016**, *219*, 57–60. [CrossRef]
60. Sak, B.; Kvác, M.; Petrzelková, K.; Kvetonová, D.; Pomajbíková, K.; Mulama, M.; Kiyang, J.; Modrý, D. Diversity of microsporidia (Fungi: Microsporidia) among captive great apes in European zoos and African sanctuaries: Evidence for zoonotic transmission? *Folia Parasitol.* **2011**, *58*, 81–86. [CrossRef]
61. Baz-González, E.; Martin-Carrillo, N.; García-Livia, K.; Abreu-Acosta, N.; Foronda, P. Molecular Detection of Microsporidia in Rabbits (*Oryctolagus cuniculus*) in Tenerife, Canary Islands, Spain. *Biology* **2022**, *11*, 1796. [CrossRef]
62. Galván, A.L.; Magnet, A.; Izquierdo, F.; Fenoy, S.; Rueda, C.; Fernández-Vadillo, C.; Henriques-Gil, N.; del Aguila, C. Molecular characterization of human-pathogenic microsporidia and *Cyclospora cayetanensis* isolated from various water sources in Spain: A year-long longitudinal study. *Appl. Environ. Microbiol.* **2013**, *79*, 449–459. [CrossRef]
63. Nogales, M.; Rodríguez-Luengo, J.L.; Marrero, P. Ecological effects and distribution of invasive non-native mammals on the Canary Islands. *Mammal Rev.* **2006**, *36*, 49–65. [CrossRef]
64. Pan, G.; Bao, J.; Ma, Z.; Song, Y.; Han, B.; Ran, M.; Li, C.; Zhou, Z. Invertebrate host responses to microsporidia infections. *Dev. Comp. Immunol.* **2018**, *83*, 104–113. [CrossRef] [PubMed]
65. Hinney, B.; Sak, B.; Joachim, A.; Kváč, M. More than a rabbit's tale—*Encephalitozoon* spp. in wild mammals and birds. *Int. J. Parasitol. Parasit. Wildl.* **2016**, *5*, 76–87. [CrossRef]
66. Medina, F.M. First record of Algerian hedgehog *Atelerix algirus* (Lereboullet, 1842) in La Palma Island biosphere reserve. *Galemys* **2016**, *28*, 61–62. [CrossRef]
67. Lobo, M.L.; Xiao, L.; Antunes, F.; Matos, O. Microsporidia as emerging pathogens and the implication for public health: A 10 year study on HIV-positive and -negative patients. *Int. J. Parasitol.* **2012**, *42*, 197–205. [CrossRef] [PubMed]

animals

MDPI

Case Report

First Report of *Sarcocystis pilosa* from a Red Fox (*Vulpes vulpes*) Released for the Re-Introduction Project in South Korea

Yeonghoon Jo [1], Sook Jin Lee [2], Mohammed Mebarek Bia [3], Seongjun Choe [3,*] and Dong-Hyuk Jeong [1,4,*]

1 Laboratory of Wildlife Medicine, College of Veterinary Medicine, Chungbuk National University, Cheongju 28644, Republic of Korea; wildwildjo@chungbuk.ac.kr
2 National Park Institute for Wildlife Conservation, Yeongju 36015, Republic of Korea; mfungus5@knps.or.kr
3 Department of Parasitology, School of Medicine and Parasite Research Center, Chungbuk National University, Cheongju 28644, Republic of Korea; biamebarek@chungbuk.ac.kr
4 The Wildlife Center of Chungbuk, Cheongju 28116, Republic of Korea
* Correspondence: parasite@chungbuk.ac.kr (S.C.); africabear@chungbuk.ac.kr (D.-H.J.)

Simple Summary: In South Korea, the red fox (*Vulpes vulpes*) has been introduced and released as part of a restoration project since 2012. One of the released individuals was found dead two months after release in 2019. During necropsy, the intestinal contents were collected and oocysts of *Sarcocystis* sp. were found during coprological examination. PCR of the 18S rRNA gene and *cox1* gene sequences was conducted, and the isolate was identified as *Sarcocystis pilosa*. This is the first report of this species in South Korea, implying that there were mammals that originally acted as intermediate and definitive hosts.

Abstract: The red fox (*Vulpes vulpes*) is a known host for various parasites, including numerous helminths and protozoans. Among these, certain species in the genus *Sarcocystis* (phylum Apicomplexa) have been documented to possess the capability to infect red foxes as definitive hosts. In South Korea, red foxes have been introduced and released as part of a re-introduction program. However, two months after its release, one of the foxes was found dead because of illegal trapping. The fox was necropsied, and a subsequent coprological study revealed oocysts of *Sarcocystis* sp. in the intestinal contents. The oocysts were identified as *Sarcocystis pilosa* based on the 18S rRNA and cytochrome *c* oxidase subunit 1 (*cox1*) gene sequences. It exhibited a 99.7–100% identity with 18S rRNA gene sequences and 99.1–99.8% identity with *cox1* gene sequences from other previously reported *S. pilosa* samples. Additionally, it showed identities of 95.4–96.4% and 91.1–91.5% with the *cox1* gene sequences of *S. hjorti* and *S. gjerdei*, while demonstrating 99.6 and 98.1% identity with the 18S rRNA gene sequences of *S. hjorti* and *S. gjerdei*, respectively. This is the first report from mainland Asia, excluding the Japanese archipelago, indicating that the life cycle of *S. pilosa* persists in South Korea.

Keywords: 18S rRNA; cytochrome *c* oxidase subunit 1; host; red fox; restoration

1. Introduction

Red foxes (*Vulpes vulpes*) have stable populations in many countries. However, interestingly, they became completely extinct in Korea in the 1990s [1]. Accordingly, a red fox restoration project was initiated in 2012 as part of an effort to establish a stable population on the Korean Peninsula [2]. These efforts may be advantageous when using foxes as sentinel species for assessing ecosystems [3]. Meanwhile, the re-introduction of foxes also presents an intricate challenge to public health, given their susceptibility to a wide range of parasites, including zoonotic parasites such as *Cryptosporidium*, *Babesia*, *Trichinella*, *Toxocara*, and *Echinococcus*, among others [3]. Among these, the genus *Sarcocystis*, a protozoan parasite classified under the phylum Apicomplexa, has a broad host range and a two-host life cycle, with specificity that varies among species. Globally, this genus includes approximately 200 recognized species [4]. In South Korea, four *Sarcocystis* species have been identified among six mammalian host species [5–10]. Within the genus *Sarcocystis*, several species (such as *S. alces*, *S. capracanis*, *S. cruzi*, *S. gracilis*, *S. hjorti*, *S. pilosa*, *S. tenella*, etc.) are recognized for utilizing canids, including red fox, as their definitive host, while concurrently relying on ruminants (cervids, cattle, sheep, etc.) as their intermediate host. This has been observed in either natural or experimental settings in previous studies conducted in U.S.A., Australia, Europe and Japan [11–15].

The majority of *Sarcocystis* spp. rely on a predator–prey relationship, in which sarcocysts formed in the muscles of intermediate hosts are ingested by definitive hosts, leading to the development of oocysts/sporocysts in the intestinal mucosa of definitive hosts and their subsequent transmission through feces. Some species have been reported to utilize humans as either intermediate or definitive hosts [16]. To elucidate and address questions related to species and host identification of *Sarcocystis* spp., molecular methods, such as the analysis of 18S rRNA and *cox1* gene sequences, have become essential. Traditional methods based on parasite morphology and intermediate host identification can be limited and inconclusive. Specifically, 18S rRNA and *cox1* genes are ideal for this purpose due to their distinct advantages. The 18S rRNA gene is highly conserved across eukaryotes, enabling robust phylogenetic reconstruction and identification of closely related species [17], while the *cox1* gene has a faster rate of evolution compared with rRNA genes, offering higher resolution within closely related species [18]. These methods offer a more precise means of understanding the relationships between different species within the same clade. When *Sarcocystis* species are clustered together in phylogenetic trees based on genetic analyses, they often share common definitive hosts. This shared evolutionary history between the parasite and its host allows for the inference of definitive hosts, even for species with known intermediate hosts but unknown definitive hosts. This inference is based on the established definitive hosts of genetically related species [19].

Sarcocystis pilosa has been reported to utilize deer (*Cervus* spp.) as an intermediate host, and is predicted to utilize Canidae as a definitive host from a phylogenetic standpoint [20]. This prediction was confirmed as a fact when *S. pilosa* was identified in the feces of foxes (*Vulpes vulpes schrencki*) from Japan [15]. However, as there are currently no reports of *S. pilosa* in Korea, the primary objective of this study is to establish foundational information for monitoring diseases and parasites during the restoration of the red fox. This will be achieved by reporting and discussing the coincidental discovery of *S. pilosa* in Korea.

2. Materials and Methods

2.1. Animal and Sample Collection

An adult female fox, coded CF1624, was initially introduced from Northeast China to South Korea as part of a restoration project in 2018, following confirmation of its genetic identity as part of a group of Northeast Asian populations. Subsequently, it was released into the wild in South Korea in 2019 after a period of acclimation. Before being released into the wild, all introduced individuals were kept in outdoor enclosures of various sizes (100–8400 m^2) in their untamed state. The foxes were provided with live chicks and chicken meat as food sources twice a day, with water provided *ad libitum*. All individuals were vaccinated for DHPPL (distemper, hepatitis, parainfluenza, parvo, and leptospirosis) and rabies and were administered anthelmintics four times during their stay in the enclosure. No significant findings were observed in pre-release fecal or blood examinations. CF1624 was found dead two months after being released because of illegal poaching. The movement of the individual was continuously monitored by GPS tracking, and the carcass was discovered a day after the movement stopped. During post-mortem examination to determine the cause of death and pathological peculiarities, the contents of the small intestine were collected and stored frozen at −20 °C. The weight of the individual before release was 6.16 kg, whereas that of the carcass was 4.84 kg.

2.2. Fecal Flotation and Cyst Observation

After fecal flotation using a saturated zinc sulfate solution, the supernatant was placed on a coverslip and observed under a light microscope (BX53; Olympus, Tokyo, Japan). Subsequently, the widths and lengths of the cystic structures were measured using the imaging software cellSens® (version 3.2, Olympus cellSens software).

2.3. DNA Extraction and PCR Sequencing

The sample, diluted with PBS buffer, underwent three cycles of treatment: 10 min in a deep freezer and 3 min at 70 °C for each cycle. Subsequently, DNA was extracted using the QIAamp PowerFecal Pro DNA Kit® (Qiagen, Hilden, Germany), following the manufacturer's instructions. The concentration of extracted DNA was measured using a NanoDrop OneC spectrophotometer® (Thermo Scientific, Waltham, MA, USA). The extracted DNA was subjected to PCR for the 18S rRNA gene (approximately 1600 bp) using the forward primer 1 L (5′-CCATGCATGTCTAAGTATAAGC-3′) [21] and the reverse primer R6 (5′-CGGAACACTCAATCGGTAGG-3′) [22] and for the cytochrome *c* oxidase subunit 1 gene (approximately 1000bp) using the forward primer SF1 (5′-ATGGCGTACAACAATCATA-AAGAA-3′) [14] and the reverse primer SR9 (5′-ATATCCATACCRCCATTGCCCAT-3′) [23], utilizing the Mastercycler Nexus gradient® (Eppendorf, Hamburg, Germany). Amplified DNA was sequenced by Cosmo Genetech (Seoul, South Korea).

2.4. Sequence and Phylogenetic Analysis

The 18S rRNA gene and *cox1* gene sequences obtained were used for phylogenetic analysis using MEGA X software version 10.0.5. For the 18S rRNA gene analysis, a set of four sequences was compiled from different geographical regions where *S. pilosa* was found, including South Korea, mainland Japan, Hokkaido, and Lithuania. Additionally, another set of 71 sequences was gathered, representing various *Sarcocystis* species and outgroup species sourced from different hosts. Notably, this compilation included sequences from *S. hjorti*, which is recognized as the species most closely related to *S. pilosa* [20].

For the analysis of the *cox1* gene, a collection of five sequences was assembled from diverse geographical locations where *S. pilosa* had been identified. These regions encompass South Korea, mainland Japan, Hokkaido, Lithuania, and Germany. Furthermore, we compiled an additional set of 35 sequences, focusing on *Sarcocystis* specimens collected from other canid fecal samples or *Sarcocystis* species recognized for utilizing canids as their definitive host. To broaden the scope, we incorporated *S. arctica* collected from canids' muscle tissue, known for utilizing avians as their definitive host [24], to serve as an outgroup species. This dataset also encompassed *S. hjorti* and *S. gjerdei*, mirroring the inclusion in the 18S rRNA gene analysis.

Sequences were sourced from GenBank and aligned using the parameters specified in the ClustalW algorithm, which was integrated into the MEGA X software. The sequences used in the present study are provided in Table S1, which can be found in the Supplementary Materials.

To facilitate further analysis, minor truncations were applied to both ends of all sequences to ensure the preservation of homologous nucleotide positions. After truncation, the phylogenetic tree dataset ultimately consisted of sequences spanning 1248 positions for the 18S rRNA gene and 1002 positions for the *cox1* gene. To construct a phylogenetic tree, we chose the most suitable substitution model for sequence evolution using the jModelTest 2.1.10. In the analysis, we employed the optimal model GTR + I + G along with its corresponding parameters. A maximum likelihood (ML) tree was constructed using the PhyML 3.1/3.0 aLRT web server (http://www.phylogeny.fr/one_task.cgi?task_type=phyml, accessed on 25 October 2023, 14 December 2023) and subsequently evaluated using the bootstrap technique with 100 iterations for bootstrapping. The ultimate tree visualization was created using FigTree version 1.4.4. 18S rRNA gene sequences of *Sarcocystis pilosa* from mainland Japan, Hokkaido, and Lithuania, as well as *S. hjorti* and *S. gjerdei* sequences, were individually compared with sequences from South Korea. A similar comparison was carried out with *cox1* gene sequences. This was accomplished using the distance calculation function in MEGA X.

3. Results

3.1. Morphological Observations

The cystic structures observed were *Sarcocystis* sporocysts, which had four sporozoites and were 16.0 (12.9~17.9) μm (SD = 0.9) long and 9.8 (7.0~13.8) μm (SD = 1.4) wide (mean values for 39 sporocysts). Some structures showed intact oocyst forms, which were composed of two sporocysts and a thin cyst wall (Figure 1).

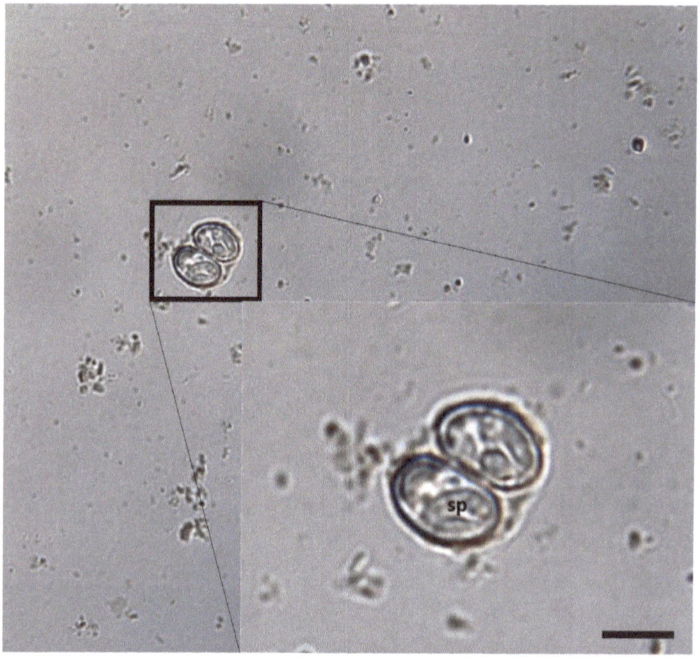

Figure 1. Morphology of a detected sporulated oocyst extracted from the feces of *Vulpes vulpes* under light microscopy. Scale bar: 10 μm; sp = sporozoite.

3.2. Molecular Characteristics

The newly generated 18S rRNA gene sequence was 1095 bp in length and was submitted to GenBank (accession number: OR724702). The sequence exhibited 100% similarity to *S. pilosa* isolated from diaphragm muscles of Lithuanian sika deer (KU753891). Furthermore, the sequence displayed 100%, 100% and 99.7% similarity with samples obtained from diaphragm muscles of Hokkaido sika deer (LC466178), feces of Hokkaido red fox (LC496069) and muscles of sika deer in mainland Japan (LC349474), respectively. The sequences were 99.6 and 98.1% identical to the 18S rRNA gene sequences of *S. hjorti* (EU282017) and *S. gjerdei* (LC349475).

The newly generated *cox1* gene sequence was 1035 bp in length and was submitted to GenBank (accession number: OR947924). The acquired sequence showed a 99.7% similarity to *S. pilosa* identified in the diaphragm muscles of Lithuanian sika deer (KU753910) and a 99.8% similarity to that identified in the diaphragm muscles of German sika deer (OP617449). Additionally, it exhibited 99.8%, 99.7% and 99.1% similarity with specimens collected from Hokkaido (diaphragm muscles of sika deer: LC466201 and feces of red fox: LC496070) and mainland Japan (muscles of sika deer: LC349967), respectively. Furthermore, the sequences showed 95.4–96.4% and 91.1–91.5% identity with the *cox1* gene sequences of *S. hjorti* and *S. gjerdei*, respectively.

3.3. Phylogenetic Analysis

The acquired 18S rRNA gene sequence formed a monophyletic group with previously reported 18S rRNA gene sequences of *S. pilosa* in the phylogenetic tree (Figure 2). Nevertheless, this monophyletic group also encompasses *S. gjerdei* and *S. hjorti*, making it challenging to achieve a complete differentiation from *S. pilosa*.

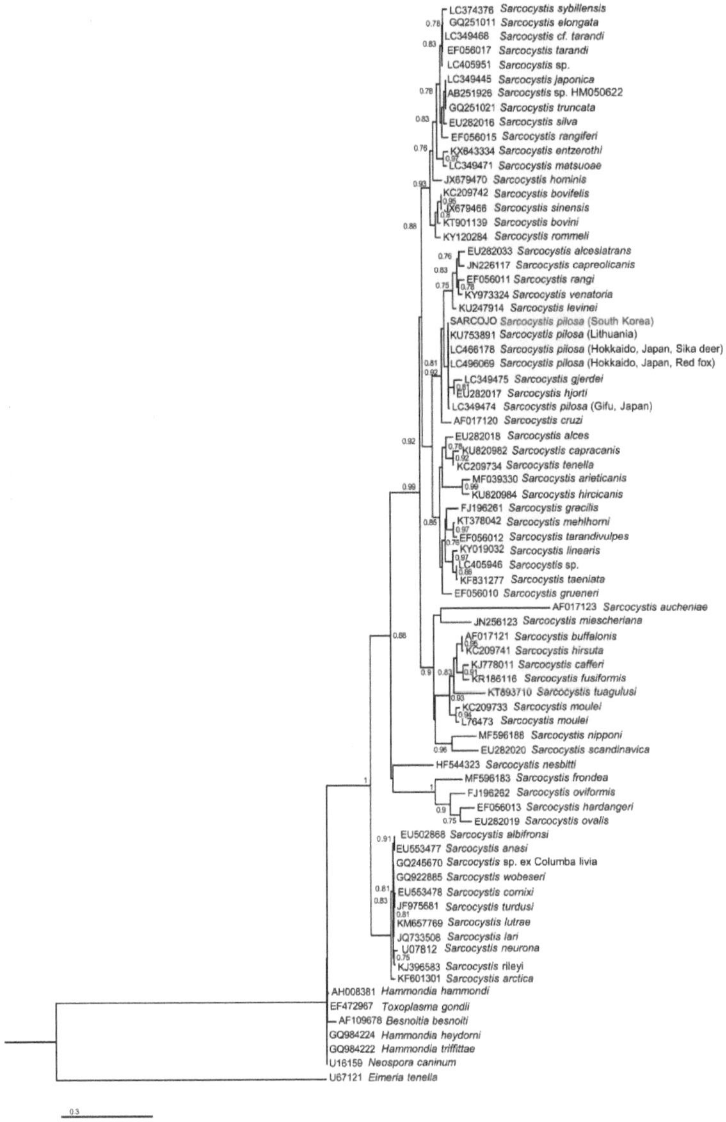

Figure 2. Maximum likelihood tree for *Sarcocystis* based on 18S rRNA gene sequences constructed using the GTR + I + G model. Bootstrap scores are expressed as proportion of 100 replications and are shown on each node. The sequence from the present study is highlighted in red. Node values less than 0.7 are not shown.

In Figure 3, the acquired *cox1* gene sequence formed a monophyletic group alongside previously reported *cox1* gene sequences of *S. pilosa*. Outgroup species (*Eimeria tenella*, *Sarcocystis arctica*) were excluded from the phylogenetic tree due to their considerable genetic divergence. The original tree is available in Figure S1 of the Supplementary Materials.

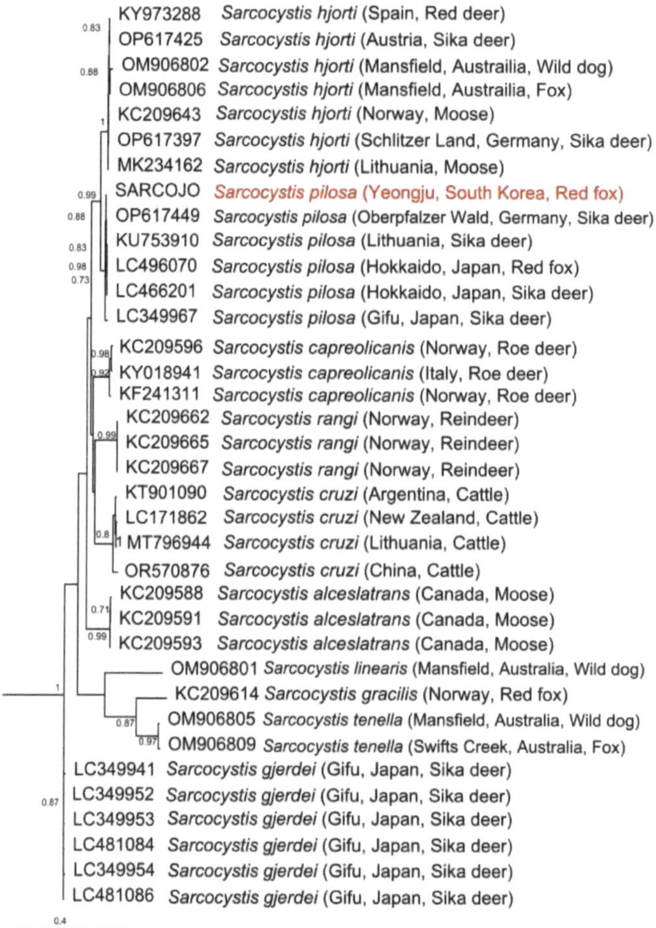

Figure 3. Maximum likelihood tree for *Sarcocystis* based on *cox1* gene sequences constructed using the GTR + I + G model. Bootstrap scores are expressed as proportion of 100 replications and are shown on each node. The sequence from the present study is highlighted in red. Node values less than 0.7 are not shown.

The distinction between *S. pilosa* and *S. hjorti* is more effectively accomplished through *cox1* gene sequences, as demonstrated in Prakas et al.'s study [20], rather than using 18S rRNA gene sequences.

4. Discussion

Sarcocystis infections such as *S. cruzi*, *S. miescheriana*, *S. tenella*, and *S. grueneri* have been reported from cattle (*Bos taurus coreanae*), pigs (*Sus scrofa domesticus*), goats (*Capra hircus coreanae*), red deer (*Cervus elaphus*), Korean water deer (*Hydropotes inermis argyropus*), and striped field mice (*Apodemus agrarius*) in South Korea [5–10]. These infected animals are intermediate hosts, and no natural infection with *Sarcocystis* has been reported in definitive hosts in South Korea, except for experimental infections conducted in dogs with *S. cruzi* [6]. In the present study, we confirmed that *Sarcocystis* sp. naturally infect red foxes in South Korea. Based on the 18S rRNA sequence, it was identified as *S. pilosa*. This is the first report of its definitive host on the main landmass of Eurasian continent. Previously reported definitive hosts of *S. pilosa* have been limited to a single fox species in Hokkaido, Japan, as in the present study. *Sarcocystis pilosa* has been previously documented in intermediate hosts such as sika deer (*Cervus nippon*) and red deer (*C. elaphus*), with confirmed occurrences in Lithuania, Japan, Switzerland, and Germany [20,25–27]. In the constructed phylogenetic tree, we confirmed that the sequences obtained in this study formed a clade with previously reported sequences of *S. pilosa*. As this was not a full-sequence analysis, it was challenging to identify significant sequence differences based on geographical distribution. However, we have confirmed the genetic similarity of *S. pilosa*, indicating its possible distribution across the Eurasian region.

Sarcocystis infection in this red fox may have originated from the source population in China before its translocation to Korea, but it also could have originated from local prey sources after its release. Considering the latent period within the definitive hosts of *Sarcocystis*, it takes approximately 14 days post-infection for oocysts to become fully sporulated in closely related species, such as *S. alces* and *S. hjorti* [13]. However, regarding the shedding period (patent period) in *S. falcatula* using Virginia opossums (*Didelphis virginiana*) as definitive hosts, shedding continued until euthanasia of all infected individuals (46–200 days post-infection) [28]. Similarly, in other studies involving dogs and cats experimentally infected with *Sarcocystis* species utilizing various intermediate hosts, such as cattle, horses, pigs, and sheep, shedding persisted until euthanasia, making it challenging to precisely determine the common shedding period of *Sarcocystis* [29]. Therefore, while it is not possible to completely exclude either of the two hypotheses, it is noteworthy that when *Sarcocystis* sp. utilizing guanacos (*Lama guanicoe*) as an intermediate host was experimentally induced in dogs, an average patent period of 45.6 days (19–61 days) was observed [30]. Additionally, experimental infections with *S. tenella* in dogs and red foxes have been reported to result in sporocyst counts of 100 or fewer per gram of feces at 60 days post-infection [12]. Considering the absence of significant findings related with internal parasite infection in the pre-release fecal examination (microscopical observation after flotation and sedimentation method) and release one year after introduction from China, it appears unlikely that the individual was infected at the time of introduction. Additionally, considering the spectrum of intermediate hosts for *S. pilosa* and closely related species, under the assumption of post-release infection from local prey sources, the reported intermediate hosts were limited to cervids, with no records beyond that.

Four native cervid species naturally coexist on the Korean Peninsula: roe deer (*Capreolus pygargus*), musk deer (*Moschus moschiferus*), sika deer (*Cervus nippon*), and water deer (*H. inermis argyropus*). Among these, the sika deer is the sole reported intermediate host of *S. pilosa*. Indeed, foxes have been observed opportunistically scavenging on cervid carcasses as a food source [31], and beyond the sika deer, other cervid species, like water deer, also hold the potential to act as intermediate hosts. Therefore, there is a need to confirm the presence of *S. pilosa* from these hosts. Furthermore, considering the proximity of the dead fox's activity area to deer farms where captive cervids (*Cervus nippon*, *C. elaphus*, *C. canadensis*) are raised, it is essential to investigate the potential infection of captive deer by *Sarcocystis* species. Additionally, it is important to ascertain which canid species served as definitive hosts before the introduction of foxes for restoration. Excluding domestic dogs, raccoon dogs (*Nyctereutes procyonoides*) are the most prevalent species among the indigenous wild canids of the Korean Peninsula. Experimental evidence has revealed its potential role as a definitive host for *Sarcocystis* species [32]. Consequently, future research employing fecal analyses should be undertaken to determine the occurrence of *Sarcocystis* infections in raccoon dogs. Such investigations would greatly aid in unraveling the ecological circulation pathways of *Sarcocystis* in domestic landscapes.

Sarcocystis infection in definitive hosts causes symptoms that are generally mild or asymptomatic and are regarded as less pathogenic. In humans, clinical symptoms such as stomach aches, nausea, and diarrhea have been reported in cases of intestinal sarcocystosis [10]. In other definitive host species, especially canines and felines, most infected individuals appear asymptomatic or may show acute-to-chronic diarrhea. However, an unusual case of megacolon and amyloidosis due to chronic inflammation induced by *Sarcocystis* sp. infection has been reported in a dog [33]. In the present study, the deceased red fox exhibited gastric perforation, intraperitoneal leakage of gastric contents, and constriction of the rectum and bladder neck. The bladder was congested with blood, and the collected intestinal contents displayed slight mucoid features. However, no histological analysis was conducted, making it difficult to correlate these findings directly with *Sarcocystis* infection.

From a One Health perspective, *Sarcocystis* typically demonstrates a generalized pattern of infecting related host species within its host range. Although the potential of *S. pilosa*, which utilizes ruminants as intermediate hosts, to infect humans seems low owing to this common pattern, it has not been experimentally confirmed. However, when non-human *Sarcocystis* spp. sporocysts are accidentally ingested, humans can also become dead-end hosts (aberrant intermediate hosts), exhibiting possible clinical symptoms of extraintestinal sarcocystosis ranging from asymptomatic muscle cysts to a severe, sudden-onset eosinophilic myositis accompanied by systemic symptoms and blood eosinophilia [10]. Therefore, it is essential to consider concerns related to water or food contamination from wildlife feces in the context of this issue.

5. Conclusions

In the present study, we confirm that, as previously reported, red foxes can serve as definitive hosts for *S. pilosa* in their natural environment. Additionally, we report the presence of *S. pilosa* in Korea, a country that is geographically close to Japan and part of continental Asia. Furthermore, this study indirectly highlights the potential intermediate hosts of *S. pilosa* and suggests the presence of mammalian species that previously acted as definitive hosts before the introduction of foxes to the Korean Peninsula. This highlights the direction for future research, including the confirmation and reporting of existing natural hosts of *S. pilosa* in South Korea. In addition, it suggests that conservation programs to restore ecosystem health may further accelerate the spread of diseases, highlighting the need to recognize the importance of disease epidemiology and quarantine, as well as the ecological implications, in this type of re-introduction program.

Supplementary Materials: The following supporting information can be downloaded at: https://www.mdpi.com/article/10.3390/ani14010089/s1, Table S1: List of sequences included in the phylogenetic analyses with details of collection localities, host species and GenBank accession numbers; Figure S1: Original phylogenetic tree of *Sarcocystis* spp. based on the *cox1* gene.

Author Contributions: Conceptualization, S.C. and D.-H.J.; methodology, Y.J., M.M.B. and S.C.; formal analysis, Y.J. and S.C.; investigation, S.J.L. and D.-H.J.; resources, S.J.L. and D.-H.J.; writing—original draft preparation, Y.J.; writing—review and editing, Y.J., S.C. and D.-H.J.; supervision, S.C. and D.-H.J.; funding acquisition, S.C. All authors have read and agreed to the published version of the manuscript.

Funding: This work was supported by the National Research Foundation of Korea (Grant No. 2020R1C1C1013563), the National Institute of Biological Resources (NIBR), and the Ministry of Environment (MOE) of the Republic of Korea (NIBR202304201).

Institutional Review Board Statement: Not applicable.

Informed Consent Statement: Not applicable.

Data Availability Statement: Data is contained within the article and Supplementary Materials.

Acknowledgments: The authors are grateful to all supporters of the Korea National Park Institute for Wildlife Conservation for their assistance with the research progress.

Conflicts of Interest: The authors declare no conflicts of interest.

References

1. Lee, H.J.; Lee, B.K.; Kwon, G.H.; Chung, C.U. Release strategy for the Red fox (*Vulpes vulpes*) restoration project in Korea Based on Population Viability Analysis. *Korean J. Environ. Ecol.* **2013**, *27*, 417–428.
2. Yu, J.N.; Han, S.H.; Kim, B.H.; Kryukov, A.P.; Kim, S.; Lee, B.Y.; Kwak, M. Insights into Korean red fox (*Vulpes vulpes*) based on mitochondrial cytochrome b sequence variation in East Asia. *Zool. Sci.* **2012**, *29*, 753–760. [CrossRef] [PubMed]
3. Garcês, A.; Pires, I. Secrets of the astute Red Fox (*Vulpes vulpes*, Linnaeus, 1758): An inside-ecosystem secret agent serving one health. *Environments* **2021**, *8*, 103. [CrossRef]
4. Dubey, J.P.; Calero-Bernal, R.; Rosenthal, B.M.; Speer, C.A.; Fayer, R. Chapter 1—General Biology. In *Sarcocystosis of Animals and Humans*; CRC Press: Boca Raton, FL, USA, 2015; pp. 1–108.
5. Moon, M.H. *Sarcocystis* infection and identification of *Sarcocystis* species in pigs in Korea. *Korean J. Vet. Res.* **1989**, *29*, 325–331.
6. Wee, S.H.; Shin, S.S. Experimental induction of the two-host life cycle of *Sarcocystis cruzi* between dogs and Korean native calves. *Korean J. Parasitol.* **2001**, *39*, 227–232. [CrossRef]
7. Son, H.; Kim, N.; Ryu, S.; Kim, H.; Rhee, J.; Cho, J.; Park, B. Ultrastructure of *Sarcocystis grueneri*-like sarcocysts from cardiac muscle of red deer (*Cervus elaphus*) in Korea. *J. Vet. Clin.* **2009**, *26*, 595–599.
8. Kim, T.-H.; Han, J.-H.; Chang, S.-N.; Kim, D.-S.; Abdelkader, T.-S.; Seok, S.-H.; Park, J.-H.; Oh, H.-S.; Kim, J.-T.; Lee, B.-H.; et al. Detection of sarcocystic infection in a wild rodent (*Apodemus agrarius chejuensis*) captured on Jeju island. *Lab. Anim. Res.* **2011**, *27*, 357–359. [CrossRef]
9. Hong, E.J.; Sim, C.; Chae, J.S.; Kim, H.C.; Park, J.; Choi, K.S.; Yu, D.H.; Park, C.H.; Yoo, J.G.; Park, B.K. Ultrastructural and molecular identification of *Sarcocystis tenella* (Protozoa, Apicomplexa) in naturally infected Korean native goats. *Vet. Med.* **2016**, *61*, 374–381. [CrossRef]
10. Kim, H.-W.; Kim, H.-C.; Ryu, S.-Y.; Choi, K.-S.; Yu, D.-H.; Park, J.; Chae, J.-S.; Park, B.K. Molecular Identification of *Sarcocystis grueneri* in Wild Korean Water Deer (*Hydropotes inermis argyropus*). *Korean J. Parasitol.* **2018**, *56*, 129–134. [CrossRef]
11. Dubey, J.P. Experimental infections of *Sarcocystis cruzi*, *Sarcocystis tenella*, *Sarcocystis capracanis* and *Toxoplasma gondii* in red foxes (*Vulpes vulpes*). *J. Wildl. Dis.* **1983**, *19*, 200–203. [CrossRef]
12. Ford, G.E. Hosts of two canid genera, the red fox and the dog, as alternate vectors in the transmission of *Sarcocystis tenella* from sheep. *Vet. Parasitol.* **1987**, *26*, 13–20. [CrossRef] [PubMed]
13. Dahlgren, S.S.; Gjerde, B. The red fox (*Vulpes vulpes*) and the arctic fox (*Vulpes lagopus*) are definitive hosts of *Sarcocystis alces* and *Sarcocystis hjorti* from moose (*Alces alces*). *Parasitology* **2010**, *137*, 1547–1557. [CrossRef] [PubMed]
14. Gjerde, B. Phylogenetic relationships among *Sarcocystis* species in cervids, cattle and sheep inferred from the mitochondrial cytochrome c oxidase subunit I gene. *Int. J. Parasitol.* **2013**, *43*, 579–591. [CrossRef] [PubMed]
15. Irie, T.; Uraguchi, K.; Ito, T.; Yamazaki, A.; Takai, S.; Yagi, K. First report of *Sarcocystis pilosa* sporocysts in feces from red fox, *Vulpes vulpes schrencki*, in Hokkaido, Japan. *Int. J. Parasitol. Parasites Wildl.* **2020**, *11*, 29–31. [CrossRef]
16. Fayer, R.; Esposito, D.H.; Dubey, J.P. Human infections with *Sarcocystis* species. *Clin. Microbiol. Rev.* **2015**, *28*, 295–311. [CrossRef]
17. Hillis, D.M.; Dixon, M.T. Ribosomal DNA: Molecular evolution and phylogenetic inference. *Q. Rev. Biol.* **1991**, *66*, 411–453. [CrossRef]
18. Hebert, P.D.; Cywinska, A.; Ball, S.L.; DeWaard, J.R. Biological identifications through DNA barcodes. *Proc. R. Soc. Lond. Ser. B Biol. Sci.* **2003**, *270*, 313–321. [CrossRef]
19. Doležel, D.; Koudela, B.; Jirků, M.; Hypša, V.; Oborník, M.; Votýpka, J.; Modrý, D.; Šlapeta, J.R.; Lukeš, J. Phylogenetic analysis of *Sarcocystis* spp. of mammals and reptiles supports the coevolution of *Sarcocystis* spp. with their final hosts. *Int. J. Parasitol.* **1999**, *29*, 795–798. [CrossRef]
20. Prakas, P.; Butkauskas, D.; Rudaitytė, E.; Kutkienė, L.; Sruoga, A.; Pūraitė, I. Morphological and molecular characterization of *Sarcocystis taeniata* and *Sarcocystis pilosa* n. sp. from the sika deer (*Cervus nippon*) in Lithuania. *Parasitol. Res.* **2016**, *115*, 3021–3032. [CrossRef]
21. Dahlgren, S.S.; Gjerde, B. Genetic characterisation of six *Sarcocystis* species from reindeer (*Rangifer tarandus tarandus*) in Norway based on the small subunit rRNA gene. *Vet. Parasitol.* **2007**, *146*, 204–213. [CrossRef]
22. Irie, T.; Ichii, O.; Nakamura, T.; Ikeda, T.; Ito, T.; Yamazaki, A.; Takai, S.; Yagi, K. Molecular characterization of three *Sarcocystis* spp. from wild sika deer (*Cervus nippon yesoensis*) in Hokkaido, Japan. *Vet. Parasitol. Reg. Stud. Rep.* **2019**, *18*, 100327. [CrossRef] [PubMed]
23. Gjerde, B. *Sarcocystis* species in red deer revisited: With a re-description of two known species as *Sarcocystis elongata* n. sp. and *Sarcocystis truncata* n. sp. based on mitochondrial *cox1* sequences. *Parasitology* **2014**, *141*, 441–452. [CrossRef] [PubMed]
24. Máca, O.; González-Solís, D. Role of three bird species in the life cycle of two *Sarcocystis* spp. (Apicomplexa, Sarcocystidae) in the Czech Republic. *Int. J. Parasitol. Parasites Wildl.* **2022**, *17*, 133–137. [CrossRef] [PubMed]
25. Basso, W.; Rojas, C.A.A.; Buob, D.; Ruetten, M.; Deplazes, P. *Sarcocystis* infection in red deer (*Cervus elaphus*) with eosinophilic myositis/fasciitis in Switzerland and involvement of red foxes (*Vulpes vulpes*) and hunting dogs in the transmission. *Int. J. Parasitol. Parasites Wildl.* **2020**, *13*, 130–141. [CrossRef]

26. Abe, N.; Matsuo, K.; Moribe, J.; Takashima, Y.; Irie, T.; Baba, T.; Gjerde, B. Morphological and molecular characteristics of seven *Sarcocystis* species from sika deer (*Cervus nippon centralis*) in Japan, including three new species. *Int. J. Parasitol. Parasites Wildl* **2019**, *10*, 252–262. [CrossRef]
27. Prakas, P.; Rehbein, S.; Rudaitytė-Lukošienė, E.; Butkauskas, D. Molecular identification of *Sarcocystis* species in sika deer (*Cervus nippon*) of free-ranging populations in Germany and Austria. *Vet. Res. Commun.* **2023**, *47*, 2165–2171. [CrossRef]
28. Porter, R.A.; Ginn, P.E.; Dame, J.B.; Greiner, E.C. Evaluation of the shedding of *Sarcocystis falcatula* sporocysts in experimentally infected Virginia opossums (*Didelphis virginiana*). *Vet. Parasitol.* **2001**, *95*, 313–319. [CrossRef]
29. Dubey, J.P.; Streitel, R.H. Shedding of *Sarcocystis* in feces of dogs and cats fed muscles of naturally infected food animals in the midwestern United States. *J. Parasitol.* **1976**, *62*, 828–830. [CrossRef]
30. Gorman, T.R.; Alcaíno, H.A.; Muñuz, H.; Cunazza, C. *Sarcocystis* sp. in guanaco (*Lama guanicoe*) and effect of temperature on its viability. *Vet. Parasitol.* **1984**, *15*, 95–101. [CrossRef]
31. Tsukada, H.; Nonaka, N. Foraging behavior of red foxes *Vulpes vulpes schrencki* utilizing human food in the Shiretoko National Park, Hokkaido. *Mammal Study* **1996**, *21*, 137–151. [CrossRef]
32. Saito, M.; Itagaki, H. Experimental infection of raccoon dogs with *Sarcocystis cruzi* and *S. miescheriana*. *J. Vet. Med. Sci.* **1994**, *56*, 671–674. [CrossRef] [PubMed]
33. Solcan, C.; Acatrinei, D.; Floristean, V.; Solcan, G.; Şlencu, B.G.; Fântânariu, M. An unusual case of megacolon due to *Sarcocystis* spp. infection and local amyloidosis in a Husky dog. *Pak. Vet. J.* **2015**, *35*, 531–533.

MDPI AG
Grosspeteranlage 5
4052 Basel
Switzerland
Tel.: +41 61 683 77 34

Animals Editorial Office
E-mail: animals@mdpi.com
www.mdpi.com/journal/animals

www.ingramcontent.com/pod-product-compliance
Lightning Source LLC
Chambersburg PA
CBHW042010260926
51850CB00020B/2967
* 9 7 8 3 7 2 5 8 4 4 3 9 5 *